probiotics

Live bacterial supplement

GÖRAN MOLIN

Probiotics

Live bacterial supplement

Göran Molin

Cover and paintings: Susie Xu

Print: Sapfo Research AB

ISBN 978-91-989045-1-2

The author Göran Molin was born 1949 and has since 1972 been working with bacterial research in connection to food and human health. He has participated as author of 205 original scientific publications.

He fulfilled his PhD in 1976; become associate professor in food hygiene at The Swedish University of Agricultural Sciences (Uppsala) in 1977; professor in food microbiology at The National Food Administration (Uppsala) in 1982; associate professor in microbiology at Lund university (Lund) in 1984; professor in food hygiene at Lund University 2000—2014; and he has upheld a position as professor emeritus at Lund University since 2014 up to present date.

Contents

Preface

In the autumn 1997, the author started an academic course in probiotics together with Siv Ahrné at the Department of Food Technology, Nutrition and Engineering (Faculty of Engineering, Lund University). In 2004 the course was extended, and has since then been carried out once a year to the present date. The course is eligible for students in the program leading to Master of Engineering (*civilingenjör*), and for the international program for Master of Food Science.

The author and Siv Ahrné were up to 2013 in charge of the course, but in connection with my retirement, the course was first coordinated by Caroline Linninge, and since 2016 by Åsa Håkansson, who today is the principal teacher of the course.

The present textbook originates from the compendium of the course, gradually formulated and written by the author during the years 1997 to 2013.

The intention with the textbook **is not** to review the full literature of probiotics, or to give an account of all important facts about probiotics and its effect on the human body. The intention with the book is to provide the reader with a tentative knowledge in the discipline of probiotics, and a starting point for further activities regarding probiotics. And last but not least, to provide the student with the necessary knowledge required for an approval at the examination of the course in *Probiotics*.

Probiotics is all about using intact microbes to achieve health beneficial effects for the human body. Therefore, if you reader, lack basic knowledge in *microb-biology* in general, and bacteriology in particular, this textbook has an appendix with the title *Student's guide to microb-biology*; this appendix is designed to give some necessary know-how in *microb-biology*, needed for the understanding of the performance of live microorganisms.

Lund, 2023-09-19
Göran Molin

Introduction

I am the living bread that came down from heaven (John 6:51), Jesus claimed, and then he said: *Whoever feeds on my flesh and drinks my blood has eternal life, and I will rise him up on the last day* (John 6:54).

This is the belief of Christians, and a crucial point in the Christian church. Thus, the statement is linked to the faith, and should not be dissected in terms of natural science. However, the present textbook is all about natural science, that is all truths of belief are linked to evidence-based knowledge in natural sciences in general, and that of microb-biology in particular. And the concept of *living bread* has to be somewhat rephrased: Probiotics are the living food that came into being by the Evolution. Whoever eat us alive have a better quality of life.

The present textbook is focused on living beings, originating from the biological kingdom of Bacteria, and how these living organisms can be processed as food or food supplements, consumed, and what happens with the human body when these live bacteria are eaten in high numbers.

The live and kicking probiotics will for certain, not bring us eternal life, but they can nevertheless contribute to an improved physiology of the human body.

If your entrance to the natural sciences goes through physics or chemistry (or even mathematics), the science of biology may seem unprecise or even unreliable. The reason is that the starting point of biological studies are the intact, living organism, and not primarily the molecular constituents of the cell.

In studies of living organism, the chemical language is not fully sufficient due to the complexity of a living organism, and the fact that a living organism is so much more than the sum of its chemical constituents. In contrast to the individual constituents (the different chemical building elements of the organism), an organism is alive.

Life drastically complicates definitions, and severely obstruct predestinations. In this respect biology are more similar to the humanities than to mathematics and chemistry.

We tend to forget that natural science, in spite of the presumptions of knowing all the different constituents of a cell; and in spite of all progress in genetic and molecular technology; the human mind is not able to create a living cell from its individual molecular constituents. Life is from a purely chemical point of view, still a totally inconceivable phenomenon.

Probiotics is to a major degree about life. The living probiotic organism encounters living cells of the human body and also other living microorganisms in the mouth and in the gastro-intestinal tract. And due to this, the scientific discussion about these meetings, and upcoming cause of events between organisms of different kind must to a substantially part be carried out in a biological language instead of a chemical one.

In fact, the bacterial systematics, including the taxonomic names of different bacteria can be more important than the intercellular chemistry. And as life is a puzzling field and biology is the science of this unaccountable life, the biological facts and the scientific discussions in biology tend to be more unprecise and doubtful than the corresponding situations within the scientific disciplines of physics and chemistry.

In contrast to mathematics where there always is one correct answer to a certain question, there are in biology often several plausible answers to choose between, and the selected one is picked out by rationality and probability.

The biological thought is dependent on the characterization of the intact, living organism, which includes the organism's location in the evolution, and the scientific (taxonomic) name of the organism.

The taxonomic name of the organism is the key to all information of the organism in question, collected in the scientific literature. This is obvious when it comes to macro-organisms as animals and plants, but less self-explanatory for microbes, invisible as they are for the naked eye. But even if they are small, and unicellular, they are still intact living organisms, even if their unicellular body is a bit tiny. Nevertheless, as Paul the Apostle states: *It is sown a natural body; it is raised a spiritual body. If there is a natural body, there is also a spiritual body* (First Corinthians 15:44).

QUESTION:

— What is life?

The concept of probiotics

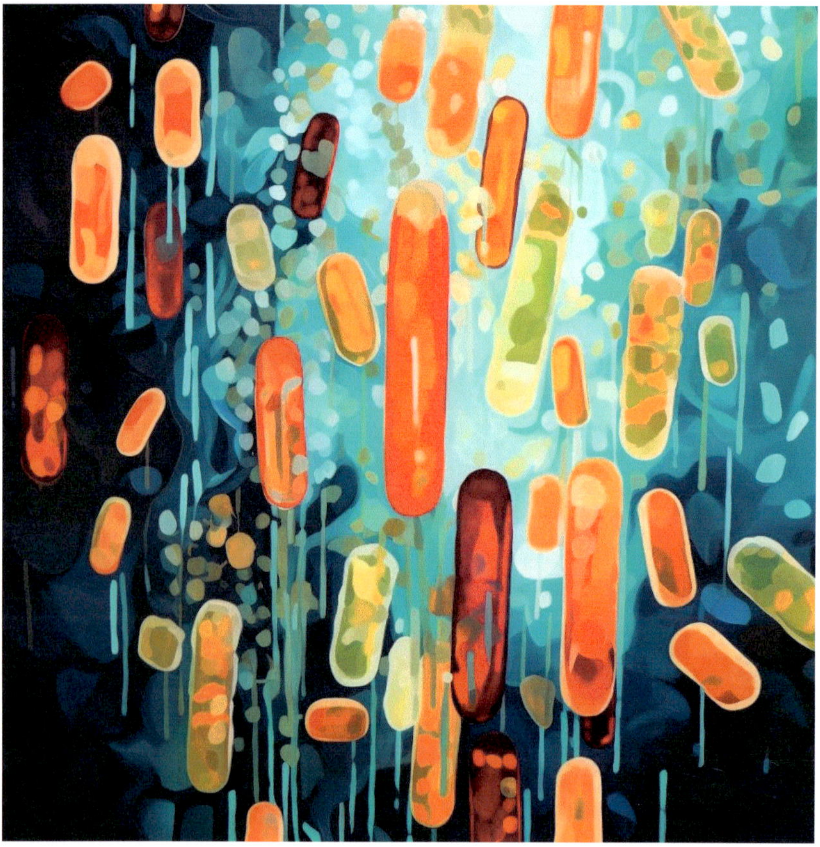

Live bacteria for ingestion

HISTORY

In the second half of the 19th Century, Louise Pasteur and Robert Koch discovered bacteria, and demonstrated that bacteria can give rise to nasty diseases, and that they also can be a major obstacle in food production and distribution as they are the main cause for much of the food spoilage. The society, including the emerging food industry, took the consequences of this and began to look upon bacteria as a major problem. The obvious solution was to eradicate all bacteria from food, for example by heat treatment. The axiom of the food industry and the health authorities became *a good bug is a dead bug*.

This is still the most popular strategy in food industry of today, in order to reduce, not only the bacterial hazards, but also to prevent unexpected quality changes during food storage.

There were early voices claiming that bacteria in spite of all adverse effects also could be beneficial. One of them was Elie Metchnikoff (1845 – 1916) who was a professor at the University of Paris, and had won the Nobel-prize 1908 for the discovery of phagocytosis.

Metchnikoff argued that certain components of the bacterial flora (microbiota) of the gut could be beneficial for the health by antagonizing harmful bacteria, that is, bacteria with physiopathological effects. The idea was published 1901 in a paper of the *British Medical Journal*, and 1907 in a book with the title, *The Prolongation of Life* (Heinemann, London).

Metchnikoff suggested that the living bacteria in lactic acid fermented milk might have health beneficial effects after ingestion. This view later underpinned the international launch of yoghurt.

Up to around 1990, milk was the only carrier of live bacteria where the producer claimed beneficial health effects. And still, milk products as yoghurt are currently the major commercial food carrier of probiotics.

Metchnikoff argued: *The intestinal flora not only influence the outcome of an infection by a pathogen, but is also responsible for the insidiously chronic toxaemia that hastens atrophy and ageing* (1901, British Medical Journal). And Metchnikoff suggested: *The introduction of this Bulgarian clotted milk into our diet may counteract, or at least diminish, the injurious effect of the intestinal flora.* As a result of this, a dairy in Paris launched "yoghurt" for first time outside Balkan. It was a sour-milk fermented with *la Lactobacilline' à la Metchnikoff.*

This yoghurt-culture happened to consist of the bacterial species *Streptococcus thermophilus* and *Lactobacillus delbrueckii* subspecies (subsp.) *bulgaricus.*

The fermentation with strains of these two taxa are still defining yoghurt in many countries, for example

Food and Drug Administration (FDA) in United States of America (USA) defines yoghurt as ”a coagulated milk obtained by lactic acid fermentation, due to *Lactobacillus bulgaricus* and *Streptococcus thermophilus*”; in current taxonomy *L. bulgaricus* is classified as *L. delbrueckii* subspecies (subsp.) *bulgaricus*.

In the early 1900-hundreds, it was argued, perhaps unjustly, that the traditional yoghurt-culture had poor survival in the passage of the acid condition of the stomach. Therefore, it was suggested in USA in the 1930-thies that *Lactobacillus acidophilus* would be a suitable supplement to yoghurt, in order to ensure the survival of live bacteria. *L. acidophilus* was claimed to have higher resistance to low pH, and thus better chances to survive the passage of the acidic environment of the stomach.

In Japan, during the 1930-thies, the *Lacticaseibacillus* (*Lactobacillus* before 2020) *casei* strain Shirota was launched as the superior beneficial bacterium for ingestion. *L. casei* Shirota has later been reclassified as the species *Lacticaseibacillus paracasei*. The strain Shirota was isolated by the Japanese scientist Minoru Shirota and is today the probiotic ingredient of the product Yakult (Yakult), which is marketed in many countries outside Japan.

Another strain of *L. paracasei* (earlier named *L. casei*) is used in the milk-based product Actimel (Danone). This strain has in Actimel been called *L. paracasei* defencis; which is the same strain as *L. paracasei* DN-114001 (=

CNCM I-1518). The strain has also been called "immunitas". However, today the strain is named *L. casei* Danone in connection with the product Actimel.

In the 1980ties, a *Lactobacillus* strain named *Lactobacillus acidophilus* GG (=ATCC 53103) was patented for its good qualifications as a live bacterial supplement, by the scientists Sherwood Gorbach and Barry Goldin. The strain was later re-identified as belonging to the species *Lacticaseibacillus rhamnosus. L. rhamnosus* GG, and has been commercially established on the international market for several decades by the Finnish company Valio Ltd.

In Finland especially, the GG-strain has been included in different kind of yoghurt-products. Strain GG differ from more typical strains of the species *L. rhamnosus* by being unable to ferment lactose.

The strain *Limosilactobacillus* (*Lactobacillus* before 2020) *reuteri* SD2112 (=ATCC 55730; = DSM 17938) was launched in Sweden 1991 as a supplement in milk ("BRA-mjölk"). This was done in cooperation between two companies, BioGaia and "ICA gruppen" AB.

L. reuteri SD2112 (DSM 17938) is together with some other strains of *L. reuteri* (*L. reuteri* ATCC PTA 6475 and *L. reuteri* ATCC PTA 5289) internationally marketed by the Swedish company BioGaia. The strain *L. reuteri* SD2112 (=DSM 17938) was originally isolated from human breast milk.

Many *L. reuteri*-strains have the ability to produce an antimicrobial substance called "reuterin" (b-hydroxy-propionaldehyde), which is an intermediate in the glycerol dissimilation.

The strain *Lactiplantibacillus* (*Lactobacillus* before 2020) *plantarum* 299v (=DSM 9843) was launched in Sweden 1994 by two companies, Probi AB and Skånemejerier (Dairies of Skåne). *L. plantarum* 299v has since the launch been included in a drink where a liquid oats gruel fermented with *L. plantarum* 299v has been mixed with a fruit-drink. The brand name in Sweden is ProViva. The product was produced and marketed by the dairy company Skånemejerier, but is today produced by Danone.

L. plantarum 299v is marketed worldwide as live supplements, and in different forms by the company Probi AB.

Examples on some other commercialized strains of lactobacilli that is marketed as live bacterial supplements with health promoting effects are *L. plantarum* HEAL9 (=DSM 15312; Probi AB), *L. paracasei* 8700:2 (DSM 13434; Probi AB), *L. paracasei* F19 (Chr. Hansen), *Lactobacillus johnsonii* La1 (Lj1; Nestlé), *Lactobacillus acidophilus* DDS-1 (UAS Labs), *Ligilactobacillus* (before 2020, *Lactobacillus*) *salivarius* UCC118 (University College Cork) and *L. acidophilus* NCFM (Wren Laboratories Ltd).

Generally speaking, strains used in commercialized, live bacterial supplements are relatively often picked from the species:
Lactiplantibacillus plantarum,
Lacticaseibacillus paracasei,
Lacticaseibacillus rhamnosus,
Limosilactobacillus fermentum,
Limosilactobacillus reuteri,
Lactobacillus acidophilus.

Beside lactobacilli, bifidobacteria are also popular as probiotics. The idea to use bifidobacteria was introduced in the 1950ies. The background was that bifidobacteria to a high degree dominate the bacterial flora (microbiota) in faeces from breast-fed infants. This fact was taken as a sign of an innate attribute, and therefore a desirable one.

The species of bifidobacteria that frequently are discussed in connection to probiotics are:
Bifidobacterium bifidum,
Bifidobacterium longum,
Bifidobacterium infantis (reclassified as *B. longum* subsp. *infantis*),
Bifidobacterium animalis,
Bifidobacterium lactis (more correctly *B. lactis* should be named *Bifidobacterium animalis* subsp. *lactis*).

The strain *B. animalis* BB12 (Christian Hansen) is the most well-known probiotic strain of bifidobacteria.

BB12 is often designated to the species *B. lactis*, especially in commercial texts.

For the time being, the yoghurt-like product Activia (Danone) is the most widespread probiotic food with bifidobacteria on the market. Activia contains the strain *B. animalis* DN-173010 (DN probably stands for the company Danone).

Lately also probiotics targeting the mouth and the upper respiratory tract have been launched on the market. One such example is the two strains, *L. reuteri* DSM 17938 and *L. reuteri* ATCC PTA 5289 (BioGaia).

Another example on oral probiotics is two strains of *Streptococcus salivarius*, coded BLIS K12 and BLIS M18 (Probi AB). The former is claimed to "promotes a healthy oral microbiota", and the latter to "supports dental health".

Dose

Humans ingest bacteria together with food, both unintentionally and on purpose. The unintentionally eaten bacteria could have adverse effects on the health when the food is contaminated with pathogenic bacteria, or if the food is contaminated with spoilage bacteria that perhaps not are directly pathogenic but can exercise negative health effects when eaten in high numbers (food close to spoilage).

On the other hand, also beneficial bacteria, *dietary bacteria* can be eaten unintentionally when fermented foods are consumed. Especially lactic acid fermented foods can contain high amounts of live bacteria and often bacteria of the same species that are used for probiotics.

The number of live bacteria ingested by traditional fermented products depends on the type of product, for example a traditional yoghurt may contain up to 10^8 –10^9 colony forming units (CFU) per ml, and a consumption of 100 ml yoghurt will, thus give the consumer a final dose of 10^{10} – 10^{11} CFU of lactic acid bacteria (LAB); while cheese may contain around 10^7 CFU per g, and consumption of 20 g cheese will yield a dose of $2x10^8$ CFU.

It should be pointed out that several commercially available, traditionally fermented products are heat-treated after the fermentation in order to prevent uncontrolled fermentation during storage, that is, no living LAB are left in the product.

Generally, a daily dose of probiotics ought to be around 10^9 CFU (preferably 10^{10} CFU) to have a chance to impose health beneficial effects; however, there is no upper limit in numbers. The higher the daily intake is, the better are the chances to obtain desirable effects. It should be held in mind that the human digestive tract already contains at least 10^{14} CFU, and a dose of 10^9 CFU will only make up 0.001% of the total number of bacteria in the human gut.

THE PROBIOTIC DEFINITION

Originally, *probiotic* meant a substance secreted by one microorganism to stimulate another [Lilley and Stillwell (1965) *Science* 147:747]. But 1974, the expression ***probiotics*** was used for organisms, or substances, which contributed to intestinal microbial balance in animals [Parker (1974) *Animal Nutrition and Health* 29:4].

The first attempt for a more precise definition of probiotics was provided by Fuller [1989; *Journal of Applied Bacteriology* 66:365]: *A live microbial feed supplement which beneficially affects the host animal by improving its intestinal microbial balance.*

In 1996 a new definition was launched, now directed solely towards human consumers: *probiotics* are *living microorganisms, which upon ingestion in certain numbers, exert health benefits beyond inherent basic nutrition* [Schaafsma (1996). Significance of probiotics in human diets. SOMED, XXIst International Congress on Microbial Ecology and Disease, Paris 23-30 October. Abstract no. 117, p. 38].

Ever since, there have been several suggestions to redefine the concept of probiotics, for example: *probiotics are a microbial dietary adjuvant that beneficially affects the host physiology by modulating mucosal and systemic immunity, as well as improving nutritional and microbial balance in the intestine* [Naidu *et al.* (1999). *Crit. Rev. Food Sci. Nutr.* 38:13].

Currently a widely accepted definition is that *probiotics are live microorganisms which when administrated in adequate amounts confer a health benefit on the host* [Anon (2002). Joint FAO/WHO Working group report on drafting guidelines for the evaluation of probiotics in food, London, Ontario, Canada].

And the same definition was given 2022 by *ISAPP: International Scientific Association for Probiotics and Prebiotics.*

The designation *probiotics* is referring to a function, and not to a taxonomical specification. However, the definition implies health beneficial effects, and there must be enough scientific evidence available to back up the health statement before a bacterial strain merits the label *probiotics.*

It is for certain not enough to refer to any strain of lactic acid bacteria or bifidobacteria as *probiotics.* A scientific and ethical problem is to decide when the health-promoting evidence is adequate for supporting the health claim, without reasonable doubt.

The health promotion is the leading theme in the probiotic concept, which also signifies that no adverse effects are allowed to be linked to an intake of probiotics.

Probiotics should be seen as a food supplement and not as a medicine. Side effects can be accepted by medicines provided that the beneficial effects are important enough, but as a health-beneficial food supplement, probiotics should be free from all type of adverse activities. Hence, probiotics should not be dangerous to ingest for any category of consumer, not even if they are very young or very old, or if they are immunocompromised. In fact, the bacteria given as probiotics should be reasonably safe even if they would end up in the circulation (in the blood).

It is also important that probiotic strains don't have qualifications for promoting the spread of antibiotic resistance, or other unwanted features. Thus, probiotics should not be prone to function as vector for mobile genes.

Linked to the concept of probiotics, other concepts involving beneficial bacteria have evolved, for example *prebiotics*, *synbiotics* and *postbiotics*.

Prebiotics can be defined as: *a substrate that is selectively utilized by host microorganisms conferring a health benefit*, or more traditionally as a non-digestible food-ingredients that selectively stimulate the growth and/or activity of health beneficial components of the microbiota of the gastro-intestinal tract.

Prebiotics have often been equated with dietary fibres, but prebiotics may also be derived from polyphenols.

Commonly-studied prebiotics are inulin, fructo-oligo-saccharides (FOS), galacto-oligo-saccharides (GOS), and human-milk-oligo-saccharides (HMOs).

Synbiotics can be defined as *a mixture comprising live microorganisms and substrate(s) selectively utilized by host microorganisms that confers a health benefit on the host.*

Postbiotics can be defined as *preparation of inanimate microorganisms and/or their components that confers a health benefit on the host.* In other words, dead microorganisms are used instead of living ones, which of course is much more convenient for the producer than live ones.

A *postbiotic* preparation originates from live micro-organisms that have been killed without harming important structures of the cell, or products that are needed for the intended effect.

Questions

— What are probiotics?

— Which are the commercially most important probiotics?

— What is the recommended daily intake of probiotics?

— Why is the concept of postbiotics seductive?

Lactic acid bacteria in traditional food

LONG TRADITION

Consumption of live lactic acid bacteria (LAB) in traditional lactic acid fermented foods, has been a regular part of the human food intake for a long time. In fact, humankind has presumably used this technique from the beginning, and lactic acid fermentation as a food preserving technique might have been invented 1.5 million years ago by the early humanoids. Thus, humans have in this way consumed large numbers of live LAB throughout their entire history of existence.

Lactic acid fermentation is the simplest and if performed correctly, maybe the safest way to preserve food. Before the Industrial Revolution (before 1760) and the urbanization, fermentation was applied just as much in Europe as it still is in many rural areas of the World.

It could very well be that the human digestive tract has evolved during continuous adaption to a more or less daily intake of living LAB.

This consumption of live LAB ceased in many industrialized countries during the twentieth century, which eventually may have led to increased frequency of gastro-intestinal (GI) and immunological dysfunctions in urbanised humans.

LAB associated to plant material were probably consumed before those associated to milk. In Sweden, as in the rest of Northern Europe, lactic acid fermented vegetables have been widely consumed up to modern times. For example, a Swedish cookery book from 1755 carefully describes procedures how to make sauerkraut, fermented spinach and fermented sorrel (*Cajsa Wargs kokbok*). However, when the emerging food industry in the twentieth century gradually took over the food production and discovered new technical means to process and preserve food, the importance of lactic acid fermentation decreased in the society.

Louis Pasteur proved that milk turned sour by the activity of bacteria (1857) and Robert Koch showed that cholera was caused by bacteria (1883). Thus, live bacteria in food were started to be looked upon with suspicion, both by the industry and by the society. So, with the best of intentions industrial leaders and health authorities set out to save the consumers from bacteria, pathogenic as well as food spoiling ones. And the obvious strategy was to systematically kill all microorganisms in food under the device – *a good bug is a dead bug.*

As a result of this strategy, the assortment of fermented products on the market decreased. And even when fermentation actually is applied, the products are often heat-treated in the final step of the process in order to kill off microorganisms. Not so much out of fear for pathogens, but more for convenience and the benefits of an improved shelf-life, with conserved eating quality.

Lactic acid fermentation

The general principle of lactic acid fermentation is that a raw-material containing carbohydrates is stored under conditions restricting the access of air (oxygen). Bacteria and yeast that contaminate the raw material will spontaneously multiply, leading to decreased oxygen tension, accumulation of carbon dioxide (CO_2), and a decrease in pH. In the end, the product will be completely dominated by lactic acid bacteria (LAB), and often also by some yeast.

The final pH in lactic acid fermented products usually is between 3.5 and 4.0. The growth of most bacteria is inhibited by the low pH.

The fermented product may on the surface be covered by moulds if the surface is exposed to air.

Yeast can also grow within the product and affect the eating quality in positive or negative directions, depending on the type of product. Yeast produces ethanol which can contribute to the preservation.

Lactic acid bacteria (LAB) have inhibitory effects on several other sorts of bacteria, primarily on gram-negative bacteria as, for example, *Enterobacteriaceae* and *Pseudomonas*, but also on some gram-positive taxa as *Clostridium* and *Listeria.* Examples on taxa that occasionally can be a problem in lactic acid fermented foods, especially if the pH doesn't become low enough, are *Bacillus cereus* and *Enterococcus.*

Different inhibitory factors can be produced during growth by lactic acid bacteria (LAB), and released into the food environment:
(*i*) carboxylic acids,
(*ii*) hydrogen peroxide,
(*iii*) nitrogen oxide, and
(*iv*) bacteriocins.

The carboxylic acids are mostly lactic and acetic acid which lower the pH. But, carboxylic acids have apart from the pH-effect, antimicrobial effects by themselves, and especially in the un-dissociated form.

Many LAB over-produce hydrogen peroxide (H_2O_2) if they are exposed to oxygen. Hydrogen peroxide is an oxidative agent and have strong antimicrobial effects.

Nitrogen oxide (NO) is produced by a limited number of LAB, that is, LAB that are able to break-down the amino acid, arginine. Nitrogen oxide has strong oxidative properties which are deleterious to many microorganisms.

Bacteriocins are often, but not always, peptides with antibacterial effects. Most bacteriocins that are produced by LAB, are active against a relatively narrow spectrum of closely related organisms. Different bacteriocins are produced by different species and by different strains within the species.

A well-known bacteriocin is nisin which is produced by *Lactococcus*. Nisin has antimicrobial effects against a relatively wide range of different bacterial taxa. Nisin is, for example, produced by different strains of *Lactococcus lactis* used as starter culture in the Swedish sour milk, *filmjölk*.

Products and processes

Ancient examples

Simple techniques. The reason for letting food undergo lactic acid fermentation are different for different applications.

Primary goals with a lactic acid fermentation could be to:

(*i*) increase the shelf-life and microbial safety of the product;

(*ii*) increase the control and the safety of food processing;

(*iii*) improve the taste and the consistency of the product;

(*iv*) achieve health beneficial effects.

A simple technique for fermentation with the prime goal to protect the food from spoiling is to dig a hole in the ground, fill in the raw material, and put a cover on top; preferably with some weights in order to press out air.

Fermentation in skin bags. In regions where permafrost can make it difficult to dig pits in the ground, instead bags made out of animal skin has been used as fermenters. Such an example was described by the Swedish explorer A.E. Nordenskiöld during the expedition around Asia in the voyages through the North-East Passage (1878–1880).

The expedition was during the winter stuck in the ice and Nordenskiöld and co-workers spent the winter together with Tschuktscer people living in Siberia on the Tschuktsch peninsula along the seashore of the North polar sea.

At this time, the Tschuktscer was a primordial society of hunters and fishermen, but a major component in their diet was lactic acid fermented plants. During the summer-months they collected different kinds of plants, such as leaf from osier (*Salix*) and the green parts of the perennial plants of the genus *Rhodiola* (rhodiola and sedum are commonly called stonecrops).

After picking, the plant-material was pressed into sealskin bags; the bags were then sealed and left for spontaneous fermentation during the summer.

In the autumn, the content freeze in the form of the out-stretched bags. The frozen mass is then, during the winter cut into pieces and eaten as it is, or together with meat or fish, or it can also be used in hot soups.

Nordenskiöld speculated that the observed consumption of lactic acid fermented plants may mimic an important part of the human diet during the Stone Age.

Sour fish. In Sweden in older days, preservation of fish was often performed by burying the fish in the ground and leave it there for spontaneous fermentation. For example, freshly caught salmon was salted and buried in the riverbank. The resulting product was called *gravad lax* (buried salmon in direct translation). The name remains in the Swedish society, but is nowadays used for salted salmon that is consumed directly, without heating.

The technique to ferment fish reaches far back in the human history. Outside the small Swedish town Sölvesborg at the shore of The Baltic Sea, archaeological excavations have exhibited ancient remains of a quite large fermentation facility for fish. The remains are about 9 000 years old (in comparison: the Egyptian pyramids are only around 4 500 years old).

Nearly 60 tons of fish estimated to have been fermented in this single facility (estimated from the number of found fishbones). The fish was mostly common roach (*Rutilus rutilus*), but also pike (*Esox lucius*), perch

(*Perca fluviatilis*) and different sorts of carp (*Cyprinidae*).

The fish was packed in bags of sealskin, sealed with seal-fat, and then buried in the ground for fermentation. The fermentation was done without addition of salt, but the process seems to some extent been controlled by addition of bark of pine.

Ethiopian *kocho*. The ancient technique to dig a hole in the ground for lactic acid fermentation, is still proving its value in Ethiopia where it is used for traditional preparing and storage of *kocho*. *Kocho* is made from the starchy root crop of Ethiopian banana, also called false banana or ensete (*Ensete ventricosum*).

Traditionally *kocho* is prepared by a lactic acid fermentation (Figure 1).

The pseudo stem and corm are pulverized manually with the help of a long wooden pestle, and pounded into a pulp from which the fibres are removed. The remaining scrapings, the pulp and the inner corm, are kneaded together, rolled into balls and wrapped in *ensete*-leaves.

The leave-wrapped packages with raw *ensete*-mash are packed in a pit in the ground that has been lined with leaves. The packages are then left for pre-fermentation for 2–5 days.

After the pre-fermentation, the packages are opened and the mash is once again mixed and thoroughly kneaded, rolled into balls, wrapped in new *ensete*-leaves, and the packages are again pressed into the pit.

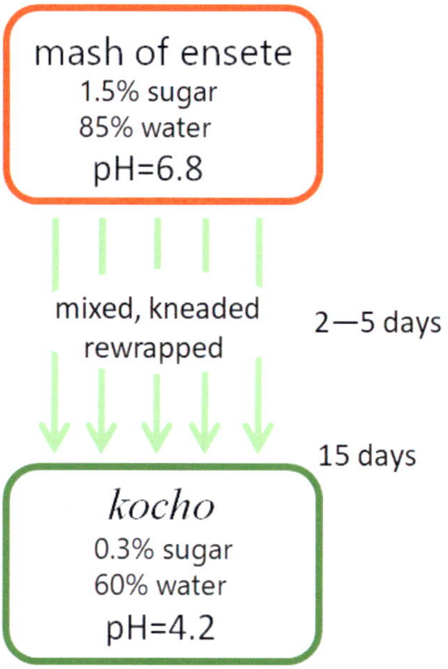

Figure 1. Flow scheme of the preparation of *kocho* from the root crop of *ensete* (*Ensete ventricosum*).

Some of the waste mash and cellulosic material from the production is put on top to create a cover, and stones are put on top with the purpose to press out air from the pit.

At a temperature of 15–18°C, a major part of the fermentation has been finalized after 2 weeks, but the *kocho* can be left in the closed pit for 6 months, up to several years.

The fermentation is initiated by *Leuconostoc* spp., but as the fermentation proceeds leuconostocs are gradually succeeded by more *Lactiplantibacillus*-like lactobacilli.

After two weeks, both lactobacilli and leconostocs are present in high numbers (about $5x10^9$ CFU per g). The viable count of lactobacilli remains high for months, while the count of leuconostocs declines.

A controlling factor is the pH as leuconsocs are less resistant to low pH than true lactobacilli. Especially resilient to low pH is *Lactiplantibacillus plantarum*, and *L. plantarum* is also frequently dominating in lactic acid fermented, plant-based products (Figure 1).

Nigerian *ogi*. This is a traditional lactic acid fermented cereal-based product from Nigeria. It is widely consumed as porridge, at breakfast, or as a cooked stiff gel called *agidi*, which is eaten together with stews, soups or fried bean cakes. As a gruel, *ogi* is also used as a weaning food.

Ogi can be made from maize, sorghum or millet, but most popular is that made from maize (*Zea mays*) or sorghum (*Sorghum bicolor* or *Sorghum dabar*). Maize is frequently used for the *ogi*-gel (agidi) while the red sorghum (*Sorghum bicolor*) often is preferred as baby-food.

Red sorghum generally gives gruel of lower viscosity than maize, which is advantageous when it is given to young children. The bulking effect of maize will reduce the intake.

Traditionally, *ogi* is manually produced by a work-intensive method (Figure 2). The cereal grains are cleaned and steeped in water for 1–3 days, when the spontaneous fermentation starts.

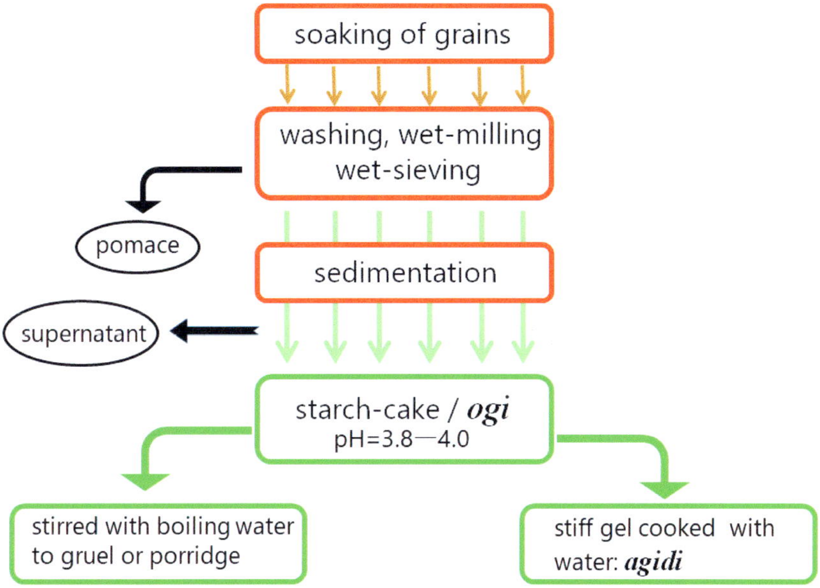

Figure 2. Flow scheme of the traditional wet-milling procedure for *ogi* production.

After the water has been poured off, the grains are wet-milled and wet-sieved through a muslin cloth or a fine wire-mesh. The pomace, mostly consisting of hulls, is discarded and can be used as animal feed.

The remaining flour suspension is left for sedimentation for 1–3 days. During this step, spontaneous lactic acid fermentation occurs.

When the *ogi* is sour enough, the supernatant is decanted and the flour cake is stirred with boiling water or with the decanted supernatant to form gruel or porridge. *Ogi* can also be cooked in water into a thick gel (*agidi*) that traditionally is put in leaf packages.

Paradoxically, in view of the fact that red sorghum is preferred in weaning food, the content of tannins is high in red sorghum, and tannins can react with proteins and make them indigestible. However, the protein digestibility of high-tannin sorghum is significantly improved by the lactic acid fermentation.

Lactiplantibacillus plantarum are able to degrade tannins and *L. plantarum* are frequently a dominating part of the microbiota in *ogi* before heating.

L. plantarum can be used as a single-strain starter for producing high quality *ogi*.

Weissella confusa can frequently be present in high numbers in spontaneously fermented *ogi*.

Tanzanian *togwa*. *Togwa* is a cereal-based, lactic acid fermented beverage, consumed in Tanzania. *Togwa* is mostly made from flour of maize (*Zea mays*), sorghum (*Sorghum bicolour*) or finger millet (*Eleusine coracana*). In some areas also rice (*Oryza sativa*) and cassava

(*Manihot esculenta*) flour, or mixtures of cassava and cereals are used. The different steps in the preparation of *togwa* are summarized in Figure 3.

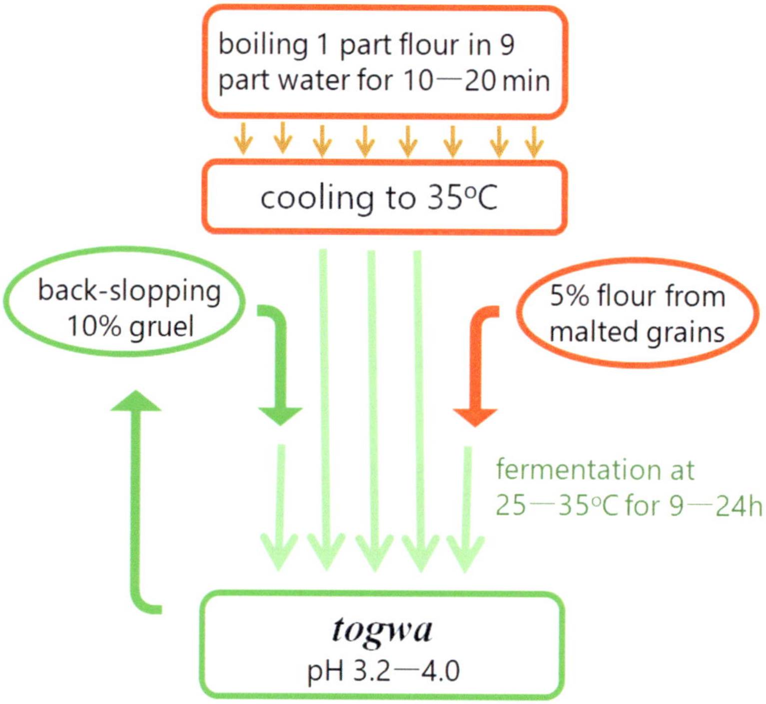

Figure 3. Flow-scheme of the production of Tanzanian *togwa*.

The flour is mixed with water (1 part flour and 9 parts of water) and cooked for 10–20 minutes. The gruel is then cooled down to about 35°C and 10% (v/v) old *togwa* is mixed into the gruel (back-slopping) together with 5% (w/v) flour made of malted grains.

Malt-flour is made from sorghum or millet. The malt-flour is prepared by soaking the grains for 12 h, then the grains are drain-dried, and spread between green leaves. A more modern alternative is to use winnowing trays, jute mats or aluminium trays and covering the grains with wet cloths.

The grains are allowed to germinate at about 30°C for 3 — 6 days until the roots and plumule become pinkish. After sun-drying and milling of the germinated grains, the malt flour is ready for use. The malting procedure reduces up to 99% of the tannins in the grains.

Heat treated gruel, malt-flour and old *togwa* (used as starter culture) are mixed, and left for 9 – 24 h (usually 10 – 12 h) for lactic acid fermentation.

Togwa has a pH of 3.2–4.0, mainly due to lactic acid formed during the fermentation. *Togwa* often contains high levels of *Lactiplantibacillus plantarum*, and *togwa* made by using *L. plantarum* as single-strain starter culture, equals spontaneously fermented *togwa* in eating quality.

By tradition, *togwa* is also given to infants as weaning food. In spite of the fact that *togwa* sometimes is produced under poor hygienic conditions, it is in the general mind known to be safe, and not linked to food-borne infections and intoxications.

Industrial applications

Salted gherkins. Cucumbers (*Cucumis sativis*) contains a high proportion of water, and are susceptible to microbial spoilage. Field-cultivated cucumbers, intended for the food industry are harvested during a short season in the autumn and are used mainly for the production of different sorts of pickles. Only a relatively small share is marketed as *salted gherkins.*

Salted gherkins is a traditional lactic acid fermented product that can be found in current supermarkets.

Field cucumbers are often heavily contaminated by microorganisms from soil. Typically, the *viable aerobic count* of bacteria can be $5x10^6$ colony forming units (CFU) per g cucumber; the viable count of *Enterobacteriaceae* can be 10^6 CFU/g, while the count of lactobacilli is fairly low, for example $5x10^3$ CFU/g. Thus, the odds for a successful preservation of cucumbers by spontaneous lactic acid fermentation seem fairly low. But by adding salt (NaCl), it becomes feasible to control the lactic acid fermentation and get a safe product with long shelf-life.

In Sweden, the cucumbers are put into a brine to achieve a final concentration of 5% NaCl. Traditionally this was done in open wood containers holding 25 tons cucumber each, and situated outdoors. After fermentation, the containers were covered with tarpaulins.

In warmer countries, a higher salt concentration is applied during fermentation (up to 8% NaCl), and after the completed fermentation, the salt concentration is increased up to 16% NaCl. Salted gherkins can be stored in the brain for at least one year.

Under the conditions that the salt concentration in the brine isn't too high, the first lactic acid bacteria (LAB) to increase in numbers are leuconostocs or related taxa, but as the fermentation proceeds, they will be succeeded by more *Lactiplantibacillus*-like lactobacilli.

A typical bacterial species found in high numbers at the end of fermentation is *Lactiplantibacillus plantarum*.

A technical problem that may occur in the fermentation of cucumbers is that gas vacuoles may build up inside the gherkins, and rise them to the surface. The phenomenon is called floating, and is caused by **malic acid fermentation**, that is malic acid in the cucumbers is catabolised to lactic acid and carbon dioxide. Many strains of *Lactiplantibacillus plantarum* possess the ability for malic acid fermentation.

Caper berries. Caper berries, or capers, are the fruits of the shrub of *Capparis*, and mainly of the species *Capparis spinosa*. The main producers of fermented capers are the Mediterranean countries such as Greece, Italy, Turkey, Morocco and Spain. The fermentation is often performed by traditional, manual procedures, also in commercial production.

The fruits are collected from the shrubs during June and/or July and are directly immersed in tap water. Here the capers spontaneously ferment for 5—7 days at a temperature range of 23—43°C. After the fermentation, the capers are put in brine to a final NaCl-concentration of about 10% (w/v).

The fermentation can, for example take place in 150 litre vats. During fermentation the pH falls below 4, and typically stops around pH 3.3.

The *total aerobic count* of the mixture of newly harvested berries directly after they have been immersed in tap water is around 10^3 CFU per ml. But after the fermentation, the total count, which in this case mainly consists of lactobacilli, rise to 10^8 CFU per ml.

The bacterial microbiota after the fermentation is typically dominated by *Lactiplantibacillus plantarum*, which can be expected to constitute around 50% of the microbiota. Other species that often occur are *Pediococcus pentosaceus*, *Levilactobacillus brevis*, *Limosilactobacillus fermentum* and *Lactiplantibacillus pentosus*.

Sauerkraut. In a Swedish cookery book by from 1755, a procedure to make sauerkraut is given:

The bottom of a tight barrel is covered with the white cabbage leaves and then it is filled with good measure of finely chopped white cabbage which gradually is pressed down with a wooden pole until the cabbage is immersed in juice. Into the barrel also a few barberry berries, some salt, a few roasted peas and dill where the seed have been carefully removed are thrown in. Some caraway is put on the top and then all is covered with cabbage leaves, and a barrel lid that can go down into the barrel is put on top and pressed down into the barrel with heavy stones. The barrel is left for fermentation in a room that is neither too hot nor too cold. When the fermentation is ended the stones are removed and a lid put on the barrel that is stored as cold as possible.

[Cajsa Wargs Kokbok, 1755]

Sauerkraut has since long a reputation of being healthy. This is reflected by the fact that Captain James Cook during his voyage around the world (1768–1780) forced his crew to eat sauerkraut. James Cook became famous, not only for his geographic discoveries but also for his extraordinary record of high survival of seamen on his ships.

An industrial example of sauerkraut production in Wisconsin (USA) is given below:

The lactic acid fermentation takes place in concrete tanks with a volume of about 100 m^3, and the fermentation is carried out in 2.3 % NaCl; the salt is added by a dry salting process of shredded cabbage.

The cabbage is manually spread in the tank, and then covered with plastic sheeting, that initially is weighted down with water on top of the plastic sheeting (Figure 4).

The temperature is left uncontrolled, but it is usually around 18°C.

Glucose and fructose which are the primary fermentable sugars in the cabbage are initially present in a concentration around 2 % (w/w).

The fermentation is run for about 14 days. Lactic acid and acetic acid are produced, and pH drops to 3.7—3.4.

Originally the cabbage has a total viable count of about $5x10^6$ CFU/g and a count of *Enterobacteriaceae* of about 10^6 CFU/g.

During the first week of fermentation, there is a rapid decrease in *Enterobacteriaceae*, and an increase in lactic acid bacteria (LAB), and after 3—7 days different taxa of lactobacilli, mainly *Weissella* and *Pediococcus* reach substantial numbers.

After 14 days, the sauerkraut is to 80—100% dominated by *Lactiplantibacillus plantarum*.

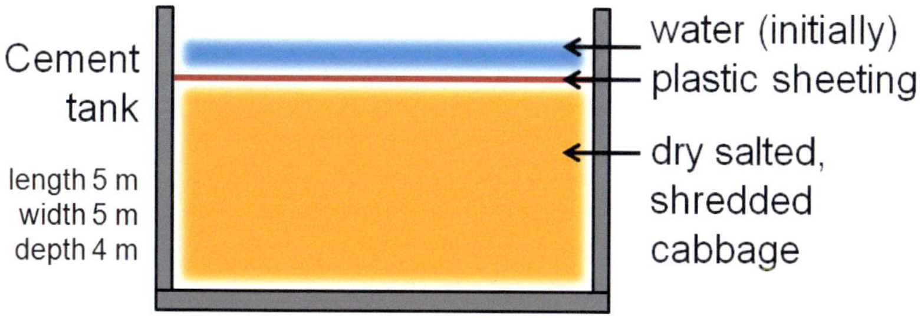

Figure 4. Tank for industrial sauerkraut production in Wisconsin (USA). [Plengvidhya *et al.* (2007). DNA fingerprinting of lactic acid bacteria in sauerkraut fermentation. *Applied and Environmental Microbiology* 73:7697]

Korean *kimchi*. A popular lactic acid fermented product is the Korean fermented food *kimchi*. *Kimchi* is processed in a similar way as sauerkraut, but it is not solely based on cabbage, but on a wide range of different kind of vegetables and spices. There are hundreds of different varieties of *kimchi* in Korea, but major ingredients are mostly Chinese cabbage and radish. Often used spices are red pepper, garlic, green onion and ginger. Salt (NaCl) is added to 2—3%, and then the mixture is fermented for 2—3 d at temperatures around 20°C.

In the beginning of the fermentation *Leuconostoc*, or similar taxa are increasing in numbers, but gradually they are outnumbered by lactobacilli as *Weissella*, *Pediococcus* and *Lactiplantibacillus*.

The load of bacteria in the final product will be around 10^8—10^9 CFU LAB per gram product, and the final pH will be 3.6—4.5.

Kimchi is increasingly known outside Korea, and is a food-concept with international potential.

Green olives in brine. The lactic acid fermentation of green olives (Spanish-style) starts with a pre-treatment with lye (1.3—2.6% NaOH [w/w]) for 5—7 hours in order to hydrolyze and remove some of the bitter tasting polyphenols.

Then, the olives are put in a brine (NaCl) and subjected to spontaneous lactic acid fermentation during 2—3 weeks, where after they reach a pH below 4.

Lactiplantibacillus plantarum is normally found in high numbers at the end of the fermentation, and in coexistence with yeast.

After the fermentation is finished, the salt in the brine is increased to 8% NaCl (w/w) to ensure the storage stability over extended time periods.

From a nutritional perspective, it should be noted that lactic acid fermentation often reduces phytates and some polyphenols which are compounds that might reduce the bioavailability of essential minerals as iron, zinc and calcium.

Phytates are reduced in lactic acid fermented products mainly due to the activity of enzymes (phytases) that are already present in the plant material, and become activated by the decreasing pH during fermentation. However, also yeast growing together with the lactobacilli produces phytases, and are theoretically able to brake-down phytates.

In contrast to phytates, polyphenols and phenolic acids can be digested directly by *Lactiplantibacillus, plantarum*, and related species, for example, *Lactiplantibacillus paraplantarum* and *Lactiplantibacillus pentosus*.

Yoghurt. Yoghurt was 1984 defined by FAO/WHO as *the coagulated milk product obtained by lactic acid fermentation through the action of Lactobacillus delbrueckii subsp. bulgaricus and Streptococcus thermophilus from milk and milk products.*

The different production-steps of yoghurt are shown in Figure 5: The high initial heating temperature will inactivate contaminating bacteria, but it will also help to improve the consistency of the final product, and make it thicker.

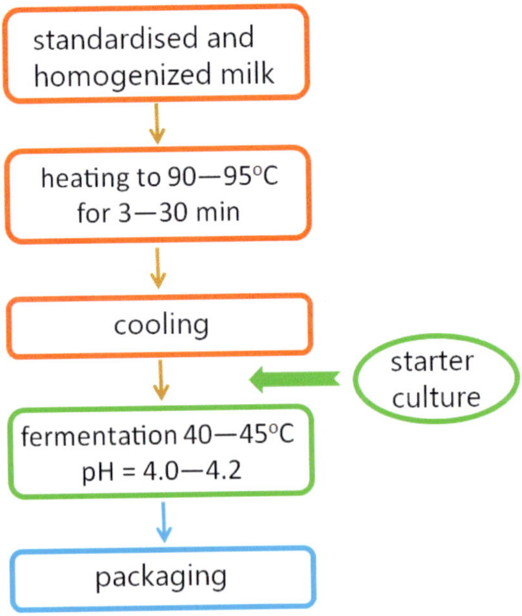

Figure 5. Flow-scheme for the production of yoghurt.

The fermentation temperature of yoghurt is relatively high (40—45°C). A consequence of the increased fermentation temperature is that the production time can be kept short (<8h), which is an advantage from an industrial, production perspective. It follows from the high fermentation temperature that also the starter culture must be able to grow at an elevated temperature.

The fermentation temperature of yoghurt has effects on the flavour. Aroma compounds that can be formed in small quantities (1—40 ppm) during fermentation are acetaldehyde, acetone, acetoin and diacetyl.

Filmjölk. The Swedish sour-milk, *filmjölk* is produced in a similar manner as yoghurt, but there are two major differences:

1) A lower fermentation temperature (19—22°C) and thereby longer fermentation time (18—20 h).

2) The use of another type of starter culture; a mixed starter culture, containing *Lactococcus lactis* and *Leuconostoc.*

The typical starter culture for *filmjölk* is not only a mixture of different species, that is: *Lactococcus lactis* subsp. *lactis*, *Lactococcus lactis* subsp. *cremoris*, *Leuconostoc mesenteroides* subsp. *cremoris* and *Leuconostoc lactis*; but a complex mixture of different, more or less undefined strains within the framework of the named species.

The same type of mixed starter culture is often used in the production of Swedish semi-hard cheese.

The milk-sugars, lactose (4.8%), glucose and galactose are fermented to lactic acid by mainly *Lactococcus lactis*.

Milk also contains citrate (1.8 g per litre), which is catabolised to acetic acid, carbon dioxide and diacetyl by *Leuconostoc* and specific strains of *Lactococcus lactis* that previously were designated *Lactococcus diacetolactis*.

Diacetyl can in turn, at least partly, be converted to acetoin. The diacetyl is given *filmjölk* much of its characteristic taste, a taste distinctly different from yoghurt.

Kefir. *Kefir* is made by fermenting milk with a mixture of yeast and lactic acid bacteria. These organisms are natural compounds in the *kefir grain* which is a spontaneously formed structure in the fermenting milk. The *kefir grain* consists of polysaccharides, proteins, yeast and bacteria.

Lentilactobacillus kefiri (former *Lactobacillus kefir*) and *Lactobacillus kefiranofaciens* are examples of lactobacilli producing the polysaccharides of the kefir grain.

During the fermentation lactic acid, ethanol and carbon dioxide are produced, together with smaller amount of diacetyl and acetoin.

Kombucha. *Kombucha* is tea (*cha* = tea) with added sugar that have been subjected to fermentation. The fermentation is performed mainly by yeast and acetic acid bacteria, for example *Gluconobacter* and *Acetobacter*. Occasionally, also lactobacilli can be involved.

The pH of *kombucha* will be around pH 2.5, which is low compared with most lactic acid fermented products. The low final pH of *kombucha* efficiently kill off most bacteria.

A *kombucha*-fermentation is started by adding so called *tea fungus*, which is a structure of polysaccharides, proteins, yeast and bacteria that spontaneously has been formed during the fermentation of *kombucha*.

Cheese. The manufacture of cheese comprises a chain of operations, which concentrate and preserve the milk by fermentation, dehydration and salting. The principal steps in the manufacture of semi-hard cheese are shown in Figure 6. It should be emphasised that each cheese variety has its own, more precise protocol of manufacturing.

Raw milk intended for cheese manufacturing is "separated" for fat-standardisation. And the number of contaminating microorganisms is reduced by pasteurisation (70—73°C for 15—20s).

"Bactufugation" or microfiltration can also be used to remove bacteria, including bacterial spores (endospores).

The reduction of the contaminating microorganisms is crucial, as the risk of potential pathogens and product spoilers in the final cheese must be kept to a minimum. It is also important that large-scale, reproducible cheese production takes place under strictly controlled conditions in order to maintain a controlled quality of the cheese.

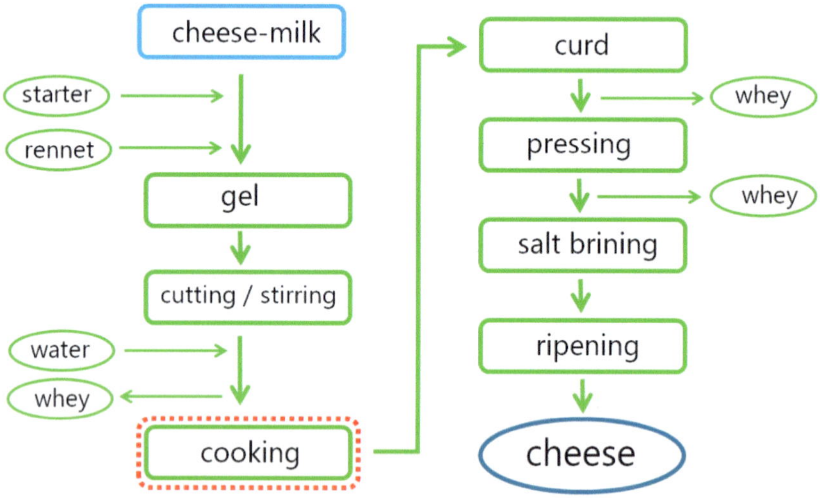

Figure 6. Flow scheme of the main steps in the manufacturing of semi-hard cheese. The "cooking" is usually performed between 38°C and 40°C during 20—40 min. The procedure up to salt brining takes 1—1.5 days. The ripening usually proceeds at about 12°C for 3—12 months.

The cheese milk is inoculated for fermentation with a starter culture. The starter cultures for Swedish, semi-hard cheese, are undefined (that is the included strains are not defined); it is so called *mesophilic LD cultures* (LD=producing L and D lactic acid), and representing several different species. The starter culture usually contains an unknown number of different strains of each species. This is regarded as a protective action against the treat of phage infections, as bacteriophages can kill the bacteria of the starter culture.

The acid producers metabolising lactose to lactic acid in the milk fermentation are strains of *Lactococcus lactis* subsp. *lactis*, and *Lactococcus lactis* subsp. *cremoris*,.

Flavour producers are *Lactococcus lactis* subsp. *lactis* bio-variant *diacetylactis*, and *Leuconostoc mesenteroides* subsp. *cremoris*, which metabolise citrate to carbon dioxide, diacetyl, acetoin and 2,3 butanediol.

The rate at which lactic acid is produced by the starter culture is controlling the water content of the cheese. The final pH of the cheese is adjusted by water addition; this is also, indirectly controlling the amount of lactose in the cheese.

The type of starter and the *cooking* temperature affect the pH of the product. Mesophilic starter cultures have optimal growth temperatures around 26°C and do not multiply at temperatures above 38—40°C. The cheese should be depleted of lactose and citrate after about one week, after the initiation of the fermentation.

Coagulation is accomplished by the addition of calf rennet, mainly the enzyme chymosin. This enzyme cleaves the κ-casein that is responsible for the micelle stability in milk, and this in combination with added $CaCl_2$ coagulates milk. When the coagulation is complete the gel is cut into cubes and converted into curd grains through the combined action of stirring, heating and decreased pH. Water control is of major importance for the final characteristics of the cheese.

The curd grains are collected in moulds, pressed to shape and then brine salted (Figure 6).

The sequential of unit operations set the conditions for the subsequent maturation during which the sensory characteristics of the cheese develop. Swedish semi-hard cheese is consumed as *mild* after approximately 3 months of ripening and as very mature (*stark* in Swedish) after about 12 months of ripening.

During the initial process which is ending when the ripening-step begins (after 1—2 days), the starter culture will have reached a level of about 10^9 CFU per gram cheese. But, during ripening the number of starter-organisms will then gradually decline.

Instead, there will be a spontaneous growth of contaminating lactic acid bacteria (LAB) during the ripening period. These bacteria are called *non-starter lactic acid bacteria* and consist in Scandinavian semi-hard cheeses, which are made from pasteurised milk, almost totally of *Lacticaseibacillus paracasei*.

In many Mediteranian cheeses, however, also *Lactiplantibacillus plantarum* is commonly occurring, especially in cheeses made from unpasteurised milk.

After about 5 months of ripening, the *non-starter lactic acid bacteria* reach substantial numbers, that is, about 10^7 to 10^8 CFU per gram cheese.

This spontaneously growing lactobacilli can affect the flavour during ripening, sometimes in a positive direction, but sometimes also in a negative way. Thus, the sensory development of the cheese is influenced by the characteristics of the dominating strain of *non-starter lactic acid bacteria*, growing and metabolising in the cheese during the ripening-period.

Fermented sausages. There are different kind of lactic acid fermented sausages with different processing times (Table 1). It is the drying of sausages that takes time, and the time is reflecting the dryness of the product.

There are two major categories of lactic acid fermented sausages:

1) Semidry sausages with a water content of 60% down to 40%, and a water activity of 0.98—0.93, respectively.

2) Dry sausages with a water activity of 40% down to 20%, and a water activity of 0.90—0.80, respectively.

Table 1. Different types of high-quality lactic acid fermented sausages.

Product	Country	Process time (d)
Medvurst	Sweden	5
Salami	Sweden	5
Salami	Italy	100
Chorizos	Portugal	5—30
Schinkenplockwurst	Germany	120

For bacteria in general, the lowest water activity for growth in food is 0.93 (some extremes can grow down to water activity 0.85), while the corresponding lowest water activity for yeast is 0.85 (some extremes can grow down to 0.80). For moulds, the corresponding limit for growth is 0.75 (some extremes down to 0.60).

Lactic acid bacteria (LAB) multiply spontaneously in the beginning of a cold smoking process, or at the initial part of a drying process. Gradually the pH of the product then drops down to 4.2—4.7 (Figure 7).

The bacterial load of the raw product after mixing of ingredients is often high (10^5—10^8 CFU per gram), while the inoculum of starter culture usually gives the mixture a concentration of 10^6 CFU starter bacteria per gram. That is, if a starter culture is added, which not always is the case.

Examples of taxa used as starter culture for lactic acid fermented sausages are *Pediococcus pentosaceus*, *Lactiplantibacillus plantarum* and *Latilactobacillus sakei*.

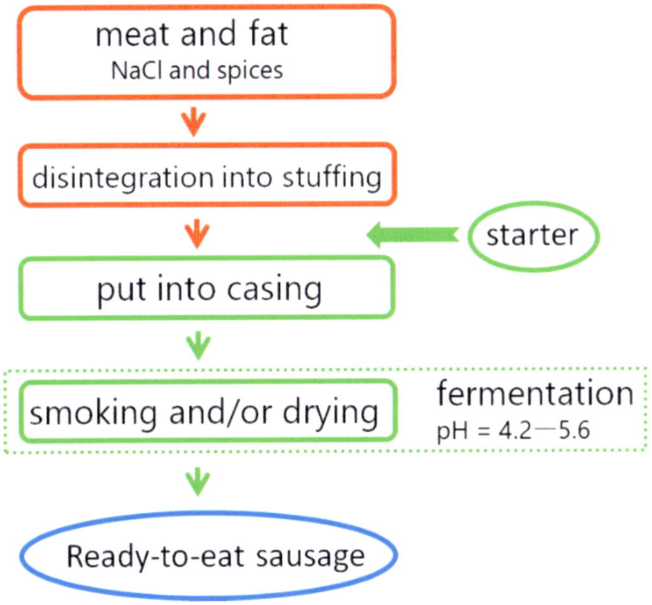

Figure 7. Flow diagram of the main steps in the manufacture of lactic acid fermented sausage.

P. pentosaceus is used as starter when the process is run at temperatures around 40—45°C, while *L. plantarum* or *L. sakei* usually are applied at temperatures ≤ 30°C which means that the product is subjected for cold smoking or drying without any smoking.

FERMENTED FOODS AS CARRIER

Empirically based applications

Existing applications of lactic acid fermented foods can be used as examples how probiotics efficiently can be integrated into food. By eating the lactic acid fermented food described in this textbook (see above), the consumer is also ingesting high numbers of live lactic acid bacteria, that have been spontaneously multiplying in the product.

We also know from empiric experience that these products are safe and so are their content of lactic acid bacteria (LAB).

On the other hand, it must be pointed out that even if these products are safe, the included bacteria do not necessarily add any health beneficial effects to the human body. All lactic acid bacteria should NOT be regarded as probiotics. However, by introducing a specific bacterial strain of LAB, with proved health beneficial abilities, into a lactic acid fermented food, the outcome can be a successful, health beneficial product concept, or in other words, a probiotic functional food concept.

Lactic acid bacteria that are found in high numbers in fermented products are by obvious reasons thriving and have good survival in the product, that is, these taxa can favourably be combined with the corresponding type of food products.

Using a probiotic strain in a food environment that is suited for its growth also means that the probiotic bacterium can be added to the food as a starter culture, and then by its own account multiply to high numbers. The commercial advantage is that it is much cheaper to let the probiotic bacterium spontaneously multiply in the product to high numbers, than supplement the product with separately produced bacteria, and maybe even freeze-dried ones, to the same numbers.

A functional advantage in letting the probiotics multiply in the carrier (food) is that the probiotics presumably are more metabolically active also when eaten.

A draw-back can be that the actively growing probiotics in a negative way change the eating quality of the food, for example, by producing carbon dioxide and acetic acid as heterofermentative lactic acid bacteria tend to do.

On the other hand, the probiotics may by interaction with food components also increase the health benefits of the actual food in itself. And as said, from the point of view of health promoting efficiency, a high number of active probiotics in the product is desirable.

Lactic acid fermented milk

The commercially most frequently used food-carrier for probiotics is yoghurt or yoghurt-like products based on milk. Some examples of these milk-based products are: Actimel (Danone), Yakult (Yakult), Lc1 (Nestlé), Gefilac (Valio) and Cultura (Arla). These types of milk-based products usually contain 10^5—10^8 CFU of the chosen probiotic strain per ml product.

The actual numbers of the probiotic organism are dependent on the hardiness of the particular strain, and how well it is adopted to the milk-environment. Strains of *Lacticaeibacillus rhamnosus* and *Lacticaseibacillus paracasei* are usually very well adopted to milk, and they will also ferment the milk if they are allowed to.

This means that they can easily reach high counts of live bacteria in the product, and they can also be assumed to have a good survival during storage.

On the other hand, their growth in the product can influence the eating quality of the final product, and this could be either a problem or an access. All depends in what way the taste or consistency is affected. The result can vary highly between strains.

The probiotic strain *Lacticaseibacillus rhamnosus* GG is somewhat atypical for the species *L. rhamnosus* as it is unable to ferment lactose.

Strains of *Lactobacillus acidophilus* are generally less adopted to the milk system, in spite of their historical record of being added to milk. Strains of *L. acidophilus* are usually unable to grow well in milk without addition of extra growth factors.

Bifidobacterium are generally more difficult to keep alive in products, and the viable count is often low or even zero in some products, in spite of claiming content of live *Bifidobacterium*.

Bifidobacterium animalis is a hardier species than most other *Bifidobacterium* spp., and is compared with most bifidobacteria relatively resistant to oxygen, and especially tough against storage is the probiotic strain *B. animalis* BB12.

Another example of a commercially successful strain is *B. animalis* DN-173010 (DN probably stands for Danone). This strain is included in the yoghurt-like product Activia, from the French company Danone.

Lactic acid fermented oats

A commercially successful, probiotic food product with the brand name ProViva was developed and launched in Sweden 1992. ProViva was from the beginning and until 2011 produced and marketed by the dairy company Skånemejerier (Malmö). Today ProViva is marketed and produced by ProViva AB (Lunnarp, Sweden), with Danone as the majority shareholder. However, the holder of the intellectual property rights of the probiotic strain included in ProViva is the public company Probi AB (Lund, Sweden).

ProViva is a relatively unorthodox food-concept, where a certain amount of a beverage of fermented oatmeal is mixed into a variety of different fruit drinks.

Lactiplantibacillus plantarum 299v is used as starter culture for the oatmeal fermentation, and provides the fruit-drinks with a high number of live bacteria.

The lactic acid fermented oatmeal beverage (gruel) is fermented with *L. plantarum* 299v in accordance with the scheme in Figure 8. And the fermented oatmeal gruel is then mixed into different types of fruit drinks. During the lactic acid fermentation *L. plantarum* 299v multiplies in the oatmeal gruel and reaches about 10^9 CFU of per ml gruel.

The lactic acid fermented oatmeal gruel is an integral part of ProViva, where 5% fermented oatmeal gruel is mixed with the different types of fruit drinks.

In the final product (ProViva), there is about $5x10^7$ CFU of viable *L. plantarum* 299v per ml of fruit drink. The product is stored in refrigeration.

By mixing in a higher proportion of the fermented oatmeal gruel into the fruit, the product became smoothie-like.

The viscosity of the oatmeal gruel in Figure 8 is lowered during the process by a supplement of malt flour (malted barley). The supplement of malt in combination with the heat treatment and the gradual decrease in pH during the lactic acid fermentation results in a decreased viscosity.

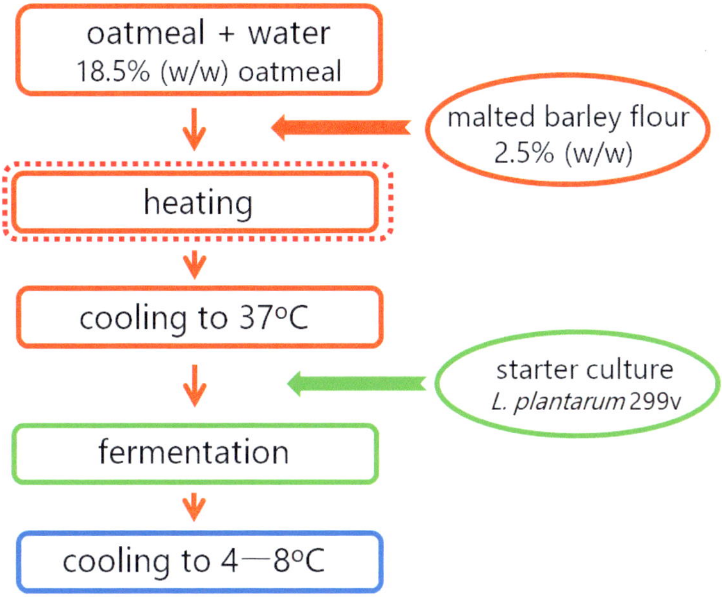

Figure 8. Flow scheme for the production of lactic acid fermented oatmeal gruel of low viscosity.

The fermented oatmeal gruel was originally intended as a base for a nutritional formula for enteral feeding (nasogastric feeding), where a liquid of low viscosity and high energy density is a prerequisite. Without added malt flour, the oat meal gruel with the stated concentration of flour (18.5%; w/w), will form a thick porridge impossible to administer through a thin tube.

The decrease in viscosity is in large part due to degradation of starch as malt is rich in amylases. There is also an increased solubility of ß-glucans which lower the viscosity. If higher amounts of malt are used, or extra malt flour is added after the heat treatment, there will be a reduction in the total amount of ß-glucans. However, the process as it is seen in Figure 8, does only cause a relatively small reduction of the total content of ß-glucans, even if the viscosity is significantly affected.

Other options for probiotic beverages

As an alternative to oats, all sorts of cereals and similar grains can advantageously be used in a probiotic food concept, that is, to use a probiotic strain as starter culture for fermentation of a gruel, and then use the fermented gruel, either more or less directly as a probiotic beverage, or mix the fermented base with other ingredients, for example juice of fruits and vegetables.

Sorghum is popular in the Tanzanian beverage *togwa* (Figure 3), which undergo a similar fermentation process as the oat-based product in Figure 8.

Dry probiotic supplements

In perspective of health-promoting efficiency, probiotics incorporated in a food environment where they thrive are presumably the best way to administrate live probiotics in high numbers. However, there are also many technical problems and marketing limitations linked to this probiotic food-concept.

An alternative is to provide the probiotic strain as a dry formula in pills, capsules, sachets and powders. The dry bacterial formula can also be incorporated in lipids and, for example be administrated in a straw.

The bacteria in these dry supplements are usually freeze dried, together with one or more protective agents.

The industrial technique to ensure a high survival of bacteria and a good shelf-life has successively improved, and most products on the current market contain a reasonably high number of live bacteria. However, the viable count of the probiotic bacteria can vary widely between different products. Preferably should all these products be stored under refrigeration as storage in room temperature substantially shorten the shelf-life based on the number of live probiotic bacteria.

Questions:

— What is the general principle for a lactic acid fermentation?

— What relevance has traditional lactic acid fermented food for probiotic food?

— Why is probiotic food based on milk the best-selling ones?

— What problems are there with lactic acid fermented foods?

Bacteria in the digestive tract

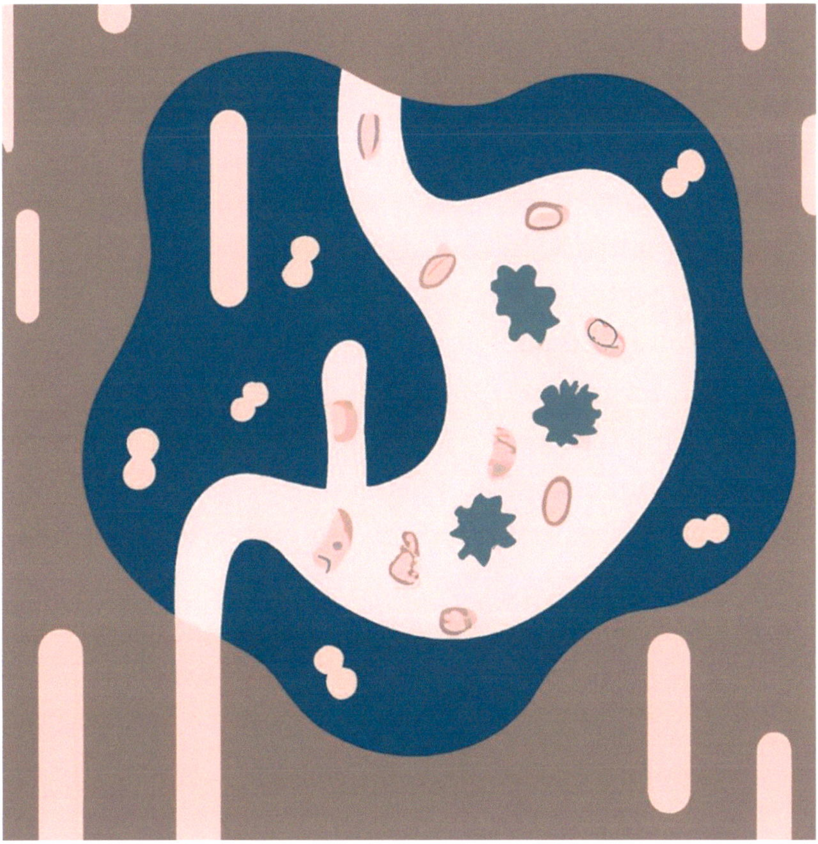

Physiology *versus* physiopathology

Humans ingest bacteria with the food, but humans also harbour huge amounts of resident bacteria (domestic bacteria) on the mucous membranes, from the mouth down to the rectum, and also in the saliva of the mouth and in the lumen of the digestive tract. Additionally, females also harbour dense populations of bacteria in the vagina.

Both ingested bacteria and resident bacteria can, depending on sort, have different effects on the human body. The bacteria can have, more or less:

(*i*) **physiological** effects, that is, effects that are consistent for normal function of the body;

or they can have:

(*ii*) **physiopathological** effects, that is, adverse effects that impose functional changes associated to injury and disease.

In other words, some resident bacteria may aggravate injuries or dysfunctions while others may work the other way around and suppress, or even have a healing effect on injuries and dysfunctions (Figure 9).

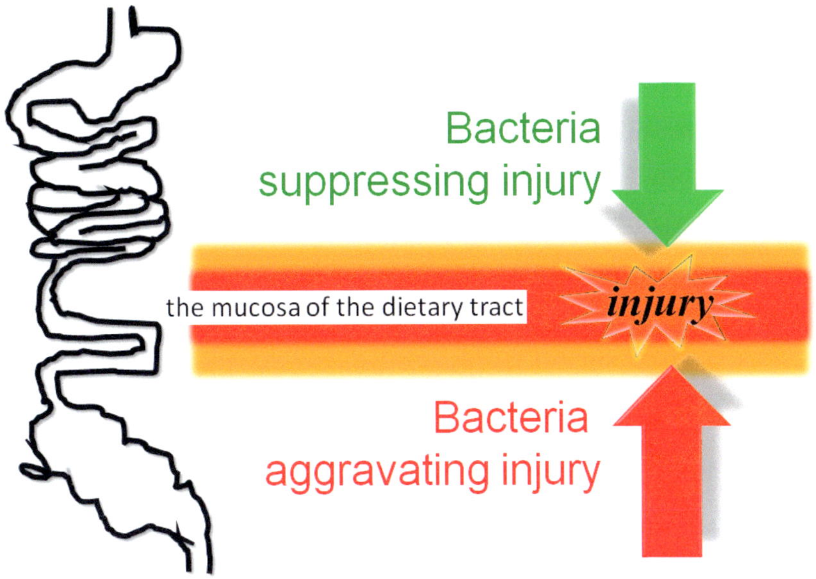

Figure 9. Schematic representation of the hypothesis that certain bacteria in the gut have physiological effects and may suppress injuries as, for example, inflammation (suppressors), while certain bacteria tend have physiopathological effects and may aggravate injuries (aggravators).

The original idea of probiotics is that by ingesting wholesome bacteria (probiotics), the beneficial bacteria of the gastro-intestinal (GI) tract can be promoted, and the balance between beneficial bacteria (suppressors) and adverse ones (aggravators) can be altered in a direction advantageous for the human health (Figure 9).

However, probiotics can do more than affecting the balance of the microbiota. Besides affect

(*i*) the composition of the resident microbiota on the mucosa and in lumen;
probiotics can affect:
(*ii*) the epithelial cells by direct cell-to-cell contact;
(*iii*) the immune system by contact with immune-competent cells (that is, macrophages, dendritic cells, B-lymphocytes, and T-lymphocytes found in Peyer's patches and other gut-associated lymphoid tissue), and antibodies;
(*iv*) the digestion by interaction with lumen-constituents (originating from both the food and the body).

Direct cell to cell-contact with the epithelial cells and with immune-competent cells, together with interaction with lumen-constituents are activities that are valid not only for probiotics but also for all the bacterial taxa making up the microbiota. Thus, by a direct intervention towards the composition of the microbiota, ingestion with probiotics may indirectly, through changes in the composition of the microbiota, lead to complex physiological effects.

The composition of the bacterial flora of the digestive tract is complex in all individuals, and this is even if only the dominating taxa are taken into account. However, a major problem when it comes to the microbiota of the GI-tract is that the physiological and physiopathological potential of most of the taxa of the microbiota still are unknown.

Exceptions are bacteria that belong to taxa with well-known pathogenic potential, or taxa with well-known, probiotic potential. These groups can with some certainty be judge as adverse or as health-beneficial, respectively.

Strains of certain species of lactobacilli and bifidobacteria have in intervention studies, in animals and in humans proved to possess beneficial health effects, and these two groups of bacteria are generally regarded as beneficial components of the human microbiota. Lactobacilli are a bit more prominent in the small intestine while bifidobacteria are more abundant in the large intestine. There is also some causal evidence of health beneficial effects of *Faecalibacterium* which mainly are resident in colon.

When it comes to adverse components of the bacterial flora of the gut, two species with well-known adverse effects are *Escherichia coli* and *Bacteroides fragilis*, which both often are abundantly found in the digestive tract. Many strains of these two species are also well-known pathogens. Even if the truly pathogenic strains not usually are consisting a normal part of the resident flora, the non-pathogenic strains of these two species have so much in common with their pathogenic relatives that they, in high numbers, are likely to impose a negative stress on the host; they are opportunistic pathogens.

Questions:

— What is implied in the concept “a balanced gut microbiota”?

— How is this concept of balance changed in the upper respiratory tract (if it is changed)?

Bacterial load

VIABLE COUNT

Human adults carry 1—2 kg bacterial biomass in the digestive tract. The human microbiota of the digestive tract starts already in the mouth, which harbours a viable count of 10^8–10^{10} colony forming units (CFU) of resident bacteria per gram saliva. These bacteria are constantly fed to the gastro-intestinal (GI) channel by the swallowing reflex.

The bacterial number is reduced in the stomach to around 10^3 CFU/g gastric juice, due to the low pH of the stomach.

The numbers remain low in the duodenum and the jejunum (10^2–10^4 CFU/g content) where there is a high intestinal motility and the flow of gastric juices are strong (bile and enzymes originating from pancreas).

The bacterial concentration increases then steeply in the ileum and colon to around 10^{10} CFU/g content in the ileum and 10^{10}–10^{12} CFU/g content in the colon. These are extremely high densities of living bacteria, and such high concentrations are seldom found in nature outside the distal part of the digestive tract of mammalians.

The bacteria of the digestive tract are of different kind and, traditional attempts to identify them have been done by pure-culture technique. Currently, methods are more directed towards direct gene-identification.

A drawback with gene-directed quantification of bacteria is that both dead and living bacteria are counted, while the traditional techniques based on viable count exclusively count living bacteria.

MICROBIOTA

Core microbiota

Some species of the microbiota of the digestive tract are found in many individuals and some are found only in few individuals. The bacterial taxa making up the dominant parts in a majority of individuals are sometimes referred to as the *core* microbiota. But, due to the huge variation between different individuals, the core of bacterial taxa, especially on lower hierarchical levels, only makes up a small proportion of the total microbiota of each individual.

Most of the core-taxa belong to the phyla:
Firmicutes (correct name [CN] is *Bacillota*) and Bacteroidetes (CN: *Bacteroidota*), and to a considerably, but smaller proportion of Actinobacteria (CN:

Actinomycetota), Proteobacteria (CN: *Pseudomonadota*) and Verrucomicrobia (CN: *Verrucomicrobiota*).

When bacteria are identified on higher hierarchical levels of taxonomy such as phylum and class, the individual differences in microbiota between persons appear to be relatively small, while the individual differences become clear at lower hierarchical levels, especially below that of genus.

However, differences between habitats within the same individual are pronounced which is reflected on the microbiota. For example, there is a significant difference in the composition of the microbiota between the oral cavity and colon, and between jejunum (upper part of the small intestine) and colon within the same individual.

The general profile of the microbiota of an individual seems to be reasonably stable over time.

The genus *Faecalibacterium*, is an example of a genus present in high levels in many individuals. *Faecalibacterium* is an efficient producer of butyric acid and is presumably high levels of *Faecalibacterium* in the gut have physiological and health beneficial effects, due to the butyric acid which provide energy to the epithelial cells of the mucosa.

The strain *Faecalibacterium prausnitzii* FBT-22 has been claimed to be a strain with probiotic potential.

Large intestine

Most data of the microbiota of the human large intestine are founded on analyses of faeces, which reasonably well reflect the colonic and rectal flora. Even if samples from the mucosa would provide a more accurate record of which bacterial taxa that actually are in direct contact with the mucosa, and thereby with the body. However, there is often a reasonably conformity between the microbiota in faeces and that on the colonic mucosa.

Generally, in humans and other mammalians, the most frequently occurring and dominating phyla of colonic bacteria are:
Firmicutes (CN: *Bacillota*; gram-positive) and
Bacteroidetes (CN: *Bacteroidota*; gram-negative).

Other frequently occurring phyla are:
Proteobacteria (CN: *Pseudomonadota*; gram-negative),
Actinobacteria (CN: *Actinomycetota*; gram-positive),
Verrucomicrobia (CN: *Verrucomicrobiota*; gram-negative),
Fusobacteria (CN: *Fusobacteriota*; atypical gram-negative) and
Spirochaetes (CN: *Spirochaetota*; gram-negative).

More recently described phyla are gradually reported to be found, but as a rule they only make up smaller proportions of the total microbiota, and occur in a more limited number of individuals.

In colon and rectum, the mucosal microbiota is often dominated by the two phyla Firmicutes (CN: *Bacillota*) and Bacteroidetetes (CN: *Bacteroidota*).

Firmicutes (CN: *Bacillota*) is represented by a wide multiple of different families and genera. But, examples of frequently reported genera belonging to Firmicutes (CN: *Bacillota*) are *Eubacterium*, *Ruminococcus*, *Clostridium* and *Faecalibacterium*.

The phylum Bacteroidetetes (CN: *Bacteroidota*) is typically represented by different species of the genera *Bacteroides* and *Prevotella*.

Less frequent but still present in considerably amounts in the large intestine are normally also the phyla Verrucomicrobia (CN: *Verrucomicrobiota*), for example represented by the genus *Akkermansia*;

Proteobacteria (*Pseudomonadota*), for example the genus *Escherichia*;

Actinobacteria (CN: *Actinomycetota*), for example the genus *Bifidobacterium*;

Fusobacteria (CN: *Fusobacteriota*), for example the genus *Fusobacterium*.

The microbiota is mostly heterogeneous and differs widely between individuals, both with regard to composition and to diversity. However, now and then a person without clinical symptoms can be grossly dominated by *E. coli* in colon, and sometimes also up in the small intestine.

E. coli must be regarded as an opportunistic pathogen, that is, an organism that can take advantage of a weakness in the hosts system of defence and become pathogenic.

Opportunistic pathogens, besides *E. coli* that can be found in high numbers in the large intestine of individuals without obvious clinical symptoms are, for example, *Bacteroides fragilis*, *Bilophila wadsworthia*, *Sutterella wadsworthensis*, *Clostridium perfringens*, *Klebsiella pneumoniae* and *Veillonella parvula*.

Oral cavity

The bacterial diversity is higher in the oral cavity than in any other part of the digestive tract, and there are huge differences between individuals.

Dominating phyla in the mouth are:
Firmicutes (CN: *Bacillota*),
Proteobacteria (CN: *Pseudomonadota*),
Actinobacteria (CN: *Actinomycetota*),
Fusobacteria (CN: *Fusobacteriota*),
Bacteroidetes (CN: *Bacteroidota*) and
Spirochaetes (CN: *Spirochaetota*).

Frequently identified genera are *Streptococcus*, *Prevotella*, *Veillonella*, *Neisseria*, *Haemophilus*, *Rothia*, *Porphyromonas* and *Fusobacterium*. The most frequently found genus in abundance, in the saliva is *Streptococcus*.

Genera as *Haemophilus*, *Neisseria*, *Streptococcus* and *Fusobacterium* include several well-known pathogenic species.

Stomach

The stomach is a harsh environment for bacteria. An adult individual produces about two litres of gastric juice daily, and therefore, the pH in lumen is below 2 under fasting conditions. However, the epithelial cells are protected by the mucus layer, and below the mucous the pH is 5—6.

Due to the low viable count of the stomach, it can always be debated whether found bacteria are resident or transient. On the other hand, *Helicobacter pylori* which is the causative agent of gastric ulcers, can definitely be regarded as resident if it can be isolated from the stomach. *H. pylori* infect the mucosa below the mucous layer.

Jejunum

The concentration of bacteria in jejunum is low. The mucosal microbiota of jejunum is typically dominated by the phylum Firmicutes, with the genus *Streptococcus* as the most abundant one.

Sometimes lactobacilli can be found as a dominating part of the microbiota.

Taxa of the phyla:
Proteobacteria (CN: *Pseudomonadota*),
Bacteroidetes (CN: *Bacteroidota*),
Fusobacteria (CN: *Fusobacteriota*) and
Actinobacteria (CN: *Actinomycetota*) can be expected to be found.

Examples of frequently found genera of Proteobacteria (CN: *Pseudomonadota*) in jejunum are: *Haemophilus*, *Escherichia*, *Acinetobacter* and *Pseudomonas*.

The phylum Bacteroidetes (*Bacteroidota*) is in jejunum often represented by the genus *Prevotella*.

Ileum

The mucosal microbiota of ileum is typically dominated by the phyla:
Firmicutes (CN: *Bacillota*) and
Bacteroidetes (*Bacteroidota*), but also
Verrucomicrobia (CN: *Verrucomicrobiota*),
Proteobacteria (*Pseudomonadota*) and
Fusobacteria (CN: *Fusobacteriota*) can be expected.

Firmicutes (CN: *Bacillota*) is in ileum usually dominated by the same family and genera as in colon.

Bacteroidetes (CN: *Bacteroidota*) is in ileum typically represented by species of the genus *Bacteroides*.

Inheritance and diet

Inheritance and diet are the two most important factors for influencing the composition the oral and intestinal microbiota of the individual. When it comes to inheritance, that is, what parents the individual originates from, it is something outside the control of the individual.

The inherent dependant differences between people are not primarily due to gender, race, or geographical origin. It has probably more to do with the mother's immunology and microbiota, that is, what the foetus has experienced during gravidity and birth. Anyway, the differences between individuals are huge.

The diet is, a factor that is open for changes, and by the diet the individual can influence the personal microbiota. Generally speaking, a vegetarian diet increases the diversity and improve the composition of the microbiota while a one-sided meat-diet decreases diversity and open up for a more adverse microbiota.

However, diets work slowly and require long-term eating for executing any measurable effects on the general composition on the microbiota.

Certain dietary components can be more beneficial for the microbiota than others. Beneficial components are, dietary fibres as for example, inulin and fructo-oligosaccharides, and certain polyphenols as the epigallo-catechin-gallat (EGCG) in tea.

Example of adverse dietary components when it comes to the microbiota are high intake of fat and of maize.

Some medicines, and especially antibiotics can be very deleterious to the composition of the microbiota.

Mother and child

Varity and diversity. Mammalians can be considered as mobile, highly efficient fermenters, producing bacteria at a high rate, and spreading them around. The "baby-fermenter" is inoculated with its first bacteria at birth, and due to the fact that the microbiota participates in shaping and tuning of the immune system of the individual, the pioneer microbiota of an infant can be of immense importance as it can be of consequence for the future immunological response and health.

The foetus has during the pregnancy only indirectly come in contact with bacteria through the blood-bound immune system of the mother. But during birth, the new-borne infant becomes directly exposed to the mother's microbiota, and this contact usually continues during the upbringing of the child. Thus, to some extent a child inherits, the microbiota of the mother, or at least the baby became contaminated with the microbiota of the mother.

The baby is exposed to bacteria originating from the mother's
(*i*) vagina,
(*ii*) rectum,
(*iii*) skin, and as soon as the mother starts to cuddle the baby, also from
(*iv*) the mouth, and if breast fed, from
(*v*) the nipples and milk-canals of the mother's breasts.

A factor overruling all other factors in determining the composition of the microbiota after a normal birth is the link to the mother. However, it is not clear if it actually is the direct transfer of bacteria from the mother that is the key factor for the establishment of the child's microbiota, or if it is the genetic and immunological inheritance from the mother that control which ones of the transferred bacteria that will become resident. When it comes to genetic inherency also (of course) the father plays his role.

The bacterial exchange with the mother can continue as a gradual long-term effect when the child lives in close contact with the mother, also after weaning, and shared diet with the mother. And now also from the father or other caregivers.

The bacterial diversity of the child's microbiota is generally increasing with age up to adulthood.

The first live bacteria that the child will meet at a normal birth, will be the vaginal bacteria of the mother. In view of the fact that a healthy vagina normally is nearly totally dominated by lactobacilli, or sometimes by bifidobacteria, and that breast milk contains prebiotic components that stimulates the growth of lactobacilli and bifidobacteria, it is reasonable to assume that lactobacilli and bifidobacteria are meant to be present in high numbers in the digestive tract of small children. And to be amongst the first bacteria to come in direct contact with the immune system along the digestive tract of the new-born baby.

Lactobacilli that most frequently dominating the vagina of fertile women are:
Lactobacillus gasseri,
Lactobacillus crispatus,
Lactobacillus iners,
Lactobacillus johnsonii, and
Lactobacillus jensenii.

Some women are dominated by *Bifidobacterium* or by a combination of lactobacilli and bifidobacteria.

The pH of the vagina is normally in the range 4—5, and the bacterial population in the vagina has a density of about 10^8 CFU per ml of vaginal secrete. Usually, one or two strains are dominating in the individual woman.

All the mentioned lactobacilli typical for the vagina can also be found in the digestive tract, but more frequently occurring, and found in higher numbers in the digestive tract are:
Lacticaseibacillus rhamnosus,
Lactiplantibacillus plantarum,
Lacticaseibacillus paracasei,
Limosilactobacillus reuteri,
Limosilactobacillus fermentum,
Ligilactobacillus salivarius, and
Lactobacillus acidophilus.

Lactobacilli in levels of 10^8 CFU per gram faeces can be found one to two days after birth in all naturally born infants (neonates). However, this early microbiota is not dominated by only lactobacilli. Usually bifidobacteia, staphylococci, *Escherichia*, *Bacteroides* amongst others can be found in higher numbers than lactobacilli.

The young child has a less diverse microbiota than the adult individual, but already after weaning, the microbiota becomes more complex, and after one year it starts to resemble an adult microbiota.

Bacterial vaginosis. The vaginal, bacterial flora can be disturbed, which leads to the syndrome of bacterial vaginosis. Bacterial vaginosis is defined as loss of lactobacilli-dominance in the vagina, and occurrence of a foul smell.

Bacterial vaginosis increases the risk for:
(*i*) urinary tract infection,
(*ii*) miss-carriage, and
(*iii*) preterm delivery.

The prevalence of bacterial vaginosis in fertile women is high in urban societies; and the risk of vaginosis increases with at high fat consumption.

Bacterial taxa that, for example, can be abundant in vagina at bacterial vaginosis are, *Prevotella*, *Atopobium*, and *Enterobacteriaceae*. These bacteria originate from the rectum.

Enterobacteriaceae and *Prevotella* have strong pro-inflammatory capabilities.

Atopic eczema. Atopic eczema is an inflammation of the skin that usually starts in early childhood. "Atopic" stands for "allergic tendencies", and children getting atopic eczema have an increased risk of developing other allergic conditions later in life, for example asthma and hay fever.

A risk-factor for getting atopic eczema is early colonization and overgrowth of *E. coli* in the digestive tract.

Overgrowth and domination with one sort of bacteria means that the bacterial diversity falls to low values, and low diversity, is by itself, increasing the risk for developing atopic eczema.

Old age

Age as such seems only to have an influence on the composition of the human GI microbiota in very young or in very old individuals. No conclusive differences in the gut microbiota can be seen between adults of different age.

However, several diseases, and medications can have effects on the microbiota of the GI-tract, and the chances for disease and heavy medication increases with age, which in turn will lower the diversity of the microbiota and increase the proportion of Proteobacteria (CN: *Pseudomonadota*). Thus, the gut microbiota in elderly individuals correlates with health status.

The diversity of the microbiota is generally lower in hospitalized individuals than in community dwellers.

Adverse bacteria

Disturbed balance. For some diseased conditions, the pathologic agent might be a disturbed microbiota, a phenomenon that also is called **dysbiosis**. Thus, the adverse effects are due to a disturbed balance of the microbiota rather than on an individual organism. The disturbed balance can also be measured as a decreased diversity of the microbiota, and as overgrowth by more aggressive fractions of resident bacteria, that is, bacteria inducing inflammatory responses in the immune system.

Species that are well-known to include pathogenic or opportunistically pathogenic strains, and that also sometimes can be found as a substantial part of the resident microbiota are *E. coli* and *Bacteroides fragilis.*

Increased proportions of *E. coli* and *B. fragilis* in the gut have been linked to inflammatory bowel disease (IBD).

Which components of the bacterial microbiota that have health beneficial effects, and which components that have health deleterious effects are a crucial question.

True pathogens as, *Salmonella* and *Shigella*, are always damaging to the body, and accordingly the immune system acts violently, and administrates a strong inflammatory response.

On the other hand, true pathogens are not present in healthy individuals. But there may be other bacteria not being clearly pathogenic but phylogenetically related to true pathogens that can be part of the resident microbiota; and these opportunistic relatives may share adverse characteristics with their pathogenic counterparts. This makes them more likely to exercise adverse effects, especially when present in high numbers.

Bacterial species that possess opportune pathogenic potential, and can make up a substantial part of the resident microbiota are, for example:
Escherichia coli (class: *Gammaproteobacteria*)
Klebsiella pneumonia (class: *Gammaproteobacteria*),
Sutterella wadsworthensis (class: *Betaproteobacteria*),
Bilophila wadsworthia (class: *Deltaproteobacteria*),
Fusobacterium varium (phylum: Fusobacteria; CN: *Fusobacteriota*),
Bacteroides fragilis (phylum: Bacteroidetes; CN: *Bacteroidota*),
Prevotella melaninogenica (phylum: Bacteroidetes; CN: *Bacteroidota*),
Vellionella parvula (phylum: Firmicutes; CN: *Bacillota*),
Streptococcus pneumonia (Firmicutes; CN: *Bacillota*),
Streptococcus anginosus (Firmicutes; CN: *Bacillota*),
Clostridium perferingens (Firmicutes; CN: *Bacillota*).

Lipopolysaccharides. Gram-negative bacteria contain lipopolysaccharide (LPS), which also is called endotoxins, as the major constituent in the outer leaflet of the outer cell membrane. LPS contains large, variable regions of polysaccharides and oligosaccharides, and a more conserved lipid region (lipid A), which is the endotoxic and biologically active moiety.

The interaction of LPS with macrophages results in the release of pro-inflammatory cytokines such as TNF-alpha, IL-6, and IL-1, and can lead to endotoxic shock, which is a fatal outcome of sepsis.

It is important for risk evaluation to know the gram-reaction of bacteria, as gram-negative bacteria contain LPS. Both the facultatively aerobic *E. coli* and the strictly anaerobic *B. fragilis* contain LPS, but the chemical structure of LPS is somewhat different between the two species. The immune system reacts differently towards the different types of LPS, and the endotoxic activity of LPS of *B. fragilis* is relatively low compared with LPS from *E. coli* and other members of the *Enterobacteriaceae* family. Nevertheless, also LPS from *Bacteroides* is a potent stimulator of the innate immune system.

Gram-negatives that typically contaminate food, and therefore are ingested on a more or less regular basis can affect both the immune system and the microbiota of the GI-tract. Typical members of the spoilage flora of many food products, and therefore also sometimes consumed in high quantities, are *Gammaproteobacteria*, for example, the families *Enterobacteriaceae* and *Pseudomonadaceae*.

Different dietary components can also affect the microbiota of the digestive tract and the response of the immune system.

A high-fat diet can increase the proportion of gram-negatives in the gut, and also increase the leakage of LPS through the intestinal barrier out into the body (Figure 10).

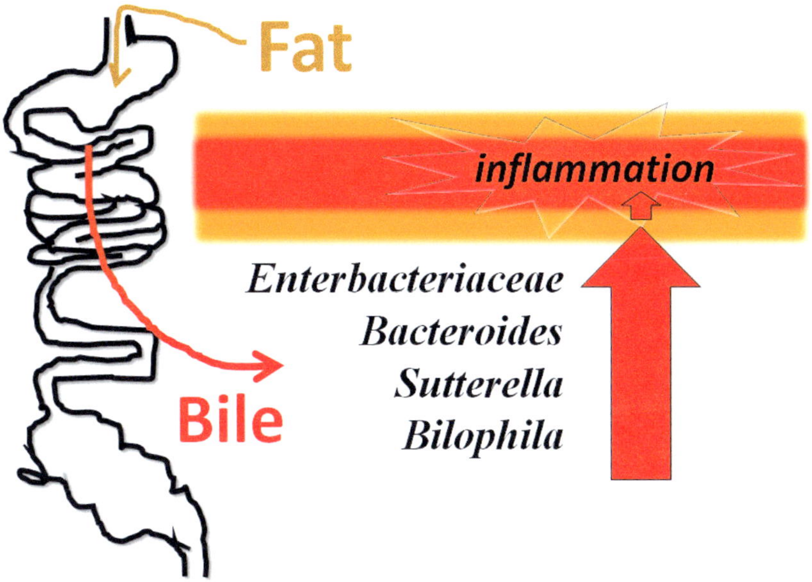

Figure 10. A high-fat diet increases the release of bile, which in turn increase the abundance of certain gram-negative species with LPS in the outer membrane, pathogenic potential, and inflammation inducing capacity. Bile is by itself a powerful detergent which can have effects on the permeability of the intestinal mucosa and, mediate increased leakage of LPS.

The reason is that a fat-rich diet increases the amount of bile in the gut, and bile has strong antimicrobial effects, but some bacterial taxa have higher resistance against bile than others. *Enterobacteriaceae* and *Bacteroides* are, for example, known for their high bile resistance.

Bile is also a powerful detergent which can have effects on the permeability of the intestinal mucosa and, thus, mediate increased leakage of LPS.

The immune system consists of different types of white blood cells and proteins.
Monocytes are big white blood cells in the blood stream which are produced by stem-cells in the bone marrow.
Macrophages which are converted monocytes are found within tissue; they perform phagocytosis.
Dendritic cells are converted monocytes that present antigens. They act as messengers between the innate and adaptive immune systems.
T-cells are a sort of "white blood cells" where the binding capacity to body-components has been evaluated by a selection process in thymus.

TNF (tumour necrosis factor) ***alpha*** is a pro-inflammatory cytokine that is produced by monocytes/macrophages and activated T-cells.
IL-1 (interleukin-1) is a group of three pro-inflammatory polypeptides (IL-1-α, IL-1β and IL-1Ra) involved in immune regulation and inflammatory response.
IL-6 (interleukin-6) initiates production of acute phase proteins and takes part in finish-off the acute inflammation (produced by macrophages and T-cells).

QUESTIONS:

— Why bother with bacterial taxonomy and identification of different bacteria when the total gene content can be sequenced?

— Can high abundance of certain bacterial taxa in the microbiota be taken as evidence for its causative effect on the host?
— If no — why not, and what information must be added?

Bacteria – body relationship

The mucous membrane

Barrier

The mucous membrane (mucosa) of the digestive tract functions as the local defence barrier that prevents the invasion and systemic spread of bacteria and other unwanted compounds, such as lipopolysaccharides (LPS).

The mucosa consists of a single layer of epithelial cells, and confers selective permeability through either a route straight through the epithelial cells (trans-cellular) or between them (paracellular).

The transfer of small molecules, for example short-chain fatty acids, amino acids, electrolytes and sugars are transporter-mediated through cells, while medium-sized hydrophilic compounds are transferred paracellularly.

Protein-sized molecules are normally prohibited from paracellular transport. The paracellular barrier function is brought about by complex structures of proteins supervising and selectively blocking the entry to the narrow gap between the epithelial cells. These gap-structures are called the **tight junctions.**

The tight junction is the closely associated area between two cells whose membranes join together. The tight junctions are responsible for the paracellular, barrier function, but the tight junctions also take part in the transfer of external stimuli to the epithelial cells, governing the proliferation and differentiation of the epithelial cells. Examples of proteins associated to the tight junction are occludin and claudin with an intra-cellular connection to zonulin.

The epithelial tissue in the mucosa of the digestive tract produces mucin that clings tightly to the epithelial surface, but mucin is also released into the lumen.

Mucins are high molecular weight, heavily glycosylated proteins that most bacteria are unable to brake-down. A key characteristic of mucins is their ability to form gels and bind to pathogens, and thus the mucins become a part of the defence system of the body, and a protective layer outside the epithelial lining.

Bacterial interaction with the epithelial cells can stimulate the mucin production of the mucosa, but the effect is highly depended on what sort of bacteria the epithelial cells are exposed for.

Pathogenic bacteria can induce mucin production, directly or indirectly, and the mucin production increases when the mucosa become inflamed. But also some beneficial bacteria can stimulate mucin production when they adhere to epithelial cells of the mucosa, without causing inflammation. Mucin-stimulating effects have been shown for lactobacilli strains, for example for the probiotic strain, *L. plantarum* 299v.

There are also bacterial groups in the microbiota of the digestive tract that has an ability to brake-down and digest mucin; such species are, *Akkermansia muciniphila* and *Ruminococcus torques*.

The genus *Akkermansia* which belongs to the phylum Verrucomicrobia (CN: *Verrucomicrobiota*) can be found in varying numbers in most adult individuals. *Akkermansia* seems to be a marker for an intestinal system in ecological balance, that is, low number of *Akkermansia* is an indication of a disturbed intestinal ecology.

It is unclear if *Akkermansia* has any health beneficial effect of its own account. On the other hand, no taxa of the phylum Verrucomicrobia (CN: *Verrucomicrobiota*) are known to be pathogenic.

LEAKAGE

Under certain conditions, the intestinal barrier function can be impaired, or overwhelmed, and allowing bacteria and lipopolysaccharides (LPS) within the digestive tract to slip through the barrier, and reach the systemic circulation, organs and tissue. This bacterial penetration is termed bacterial **translocation.**

Parts or products of bacteria are sometimes included in the expression translocation, for example bacterial components as LPS.

Bacterial translocation is the passage of viable bacteria through the epithelial mucosa into the *lamina propria* and then to the mesenteric lymph nodes and possibly other tissues.

The status of the mucosa is crucial for several categories of patients:
(*i*) patients at risk for multiple organic failure;
(*ii*) patients undergoing chemo- or radiotherapy;
(*iii*) patients with HIV;
(*iv*) patients undergoing abdominal surgery;
(*v*) patients undergoing transplantation.

There are a variety of human diseases where an abnormal permeability of the mucosa seems to be important, and at least aggravating the symptoms of the disease. Such diseases are autoimmune diseases, for example:
(*i*) type 1 diabetes,
(*ii*) celiac disease,
(*iii*) multiple sclerosis,
(*iv*) rheumatic diseases and
(*v*) Crohn's disease.

It also seems as the permeability of the intestinal mucosa is increased in atopic dermatitis and irritable bowel syndrome (IBS).

In contrast, bacterial translocation is not necessarily occurring only in connection with trauma and disease. There is evidence that LPS (endotoxins) of gut origin in minute amounts enter into the portal blood (called vein endotoxaemia), and that this is a frequently occurring phenomenon. But under normal conditions, this low-grade endotoxaemia of intestinal origin is cleared by the cells of the reticuloendothelial system of the liver.

Intestinal bacteria and bacterial products, such as LPS, reach the liver by the portal blood which is draining the digestive tract. The liver will then try to sustain normal physiology and homeostasis, and can by participating in these activities cause systemic, as well as local inflammation and other immune responses.

The liver contains parenchymal cells (hepatocytes) and non-parenchymal cells, which include endothelial cells, Kupffer cells, hepatic stellate cells and Pit cells (Pit cells are liver-specific natural killer cells).

Bacterial interaction

High amounts of LPS-synthesizing bacteria in the gut can cause low-grade systemic inflammation by a continuous leakage (translocation) of bacteria and LPS to the liver (Figure 11). This applies to gram-negative bacteria in general, but to *E. coli* and other species of the family *Enterobacteriaceae* in particular.

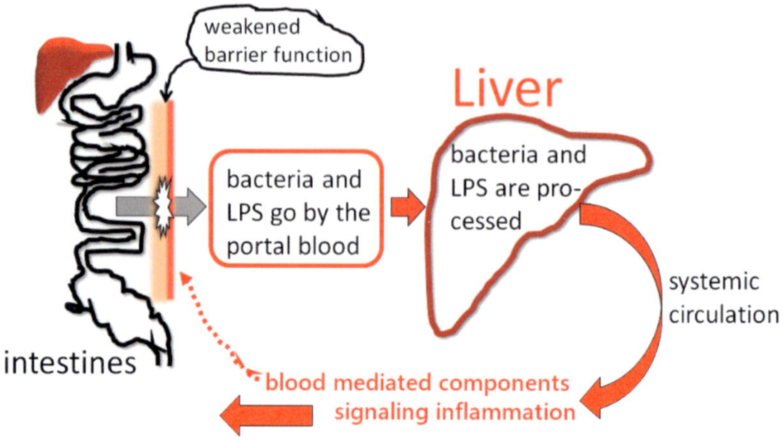

Figure 11. Translocated bacteria and bacterial components, for example, lipopolysaccharides (LPS) will by the portal blood come to the liver that will respond with inflammation in order to eliminate them.

A high proportion of *E. coli* in the human gut microbiota is not unusual for persons living in urban societies, and high loads of *E. coli* is most likely something that ought to be avoided from a health perspective.

There seems to be a link between metabolic diseases and a gut microbiota with high proportion of adverse bacterial taxa. Individuals with type 2 diabetes often have an unbalanced microbiota (**dysbiosis**) compared to healthy individuals.

Individuals with type 2 diabetes have a lower proportion of Firmicutes (CN: *Bacillota*), for example, of the genus *Faecaliebacterium*.

And a higher proportion of
Bacteroidetes (CN: *Bacteroidota*), for example of the genus *Bacteroides* and/or
Proteobacteria (*Pseudomonadota*), typically *E. coli.*

All Proteobacteria (CN: *Pseudomonadota*) have LPS in the cell-walls, and many taxa in the different classes of Proteobacteria (CN: *Pseudomonadota*) are negative for the health, and can make up harmful parts of the resident microbiota. The ingestion of probiotics, therefore become an obvious strategy to counteracting harmful Proteobacteria (CN: *Pseudomonadota*) in the digestive tract, and to get beneficial, long-term health effects as a result.

The traditional idea of probiotics presumes a balance between beneficial and adverse bacteria in the gut, and as long as the beneficial bacteria are present in adequate numbers they strengthen the gut-health.

In contrast, excessive growth of adverse bacteria creates problems. However, this indirect way of affecting the human body through the composition of the gut microbiota is not the only option for probiotics to provide health beneficial effects.

Ingested bacteria will pass through the digestive tract and hit the lining of epithelial cells covering an estimated area of about 250 m^2.

The ingested bacteria will also come into contact with immune-cells, that is, macrophages, dendritic cells, B-lymphocytes, and T-lymphocytes found in Peyer's patches and other gut-associated lymphoid tissue (GALT). These interactions between ingested probiotics, presumably starting already in the mouth, carry on through the system as the probiotics pass down the gastro-intestinal (GI) tract. Thus, ingested bacteria can trigger major effects also when passing through the upper, lowly populated parts of the GI-tract.

There are complex relationships between the intestinal immune system and the bacteria of the digestive tract. And it is crucial for the epithelial cells and the mucosal associated immune system to distinguish between pathogenic and non-pathogenic agents. Therefore, the intestinal epithelial cells are capable of detecting bacterial antigens, and initiating and regulating both innate and adaptive immune responses.

Signals from bacteria can be transmitted to adjacent immune cells such as macrophages, dendritic cells and lymphocytes through molecules expressed on the epithelial cell surface, as for example, Toll-like receptors (TLRs).

Interaction of TLRs and bacterial molecular patterns results in:

(*i*) activation of an intra-cellular signalling cascade;

(*ii*) up-regulation of inflammatory genes;

(*iii*) production of pro-inflammatory cytokines and interferons;

(*iv*) recruitment of blood cells other than lymphocytes (myeloid cells).

Interaction with TLRs also stimulates expression of co-stimulatory molecules required to induce an adaptive immune response of antigen presenting cells.

The presence of bacteria with inflammation-inducing capabilities in the digestive tract can cause an increased inflammatory tone in the body (low grade inflammation). The inflammation can be local in the mucosa, but can also involve the whole body and be systemic.

Systemic inflammation over prolonged time periods increases the risk for many diseases, for example heart and cardio-vascular diseases, diabetes type 2, non-alcoholic fatty-liver disease (NAFLD) and obesity.

More local inflammation over prolonged time periods, for example, in the tissue of the digestive tract or in the liver increases the risk for development of cancer in the inflamed site.

QUESTIONS:

— From where comes lipopolysaccharides?

— In what sense is inflammation a double-edged sword?

— How can bacteria come into contact with the interior of the body?

— Are there any bacteria in blood?

— Can gram-positive bacteria cause inflammation?

Probiotic health effects

Different types of effects

Probiotics can affect the body in different ways, and with different outcomes for the health.

One important probiotic way to affect the body in a way that is crucial for health is to improve the condition of the mucosa, and strengthen the epithelial barrier function.

A decreased barrier function of the mucosa and, consequently, an increased leakage of bacteria and other unwanted components through the mucosa is a general problem linked to several different diseases and malfunctions.

Another crucial probiotic interaction is that with the immune-system which can lead to major health consequences. This immunological interaction starts already in the mouth, but is probably most pronounced in the small intestine where the immune activities of the body is especially intense.

Probiotics can affect the health in the following general ways:
1) Probiotics can modulate the immune response in a beneficial way.

2) Probiotics can strengthen the tight-junctions between the epithelial cells in the mucous membrane, and in this way improve the barrier effect of the mucosa.

3) Probiotics can stimulate the epithelial cells of the mucous membrane to produce more mucin, and in this way improve the barrier effect of the mucosa.

4) Probiotics can decrease the proportion of adverse bacteria in the mouth and in gastro-intestinal (GI) tract.

5) Probiotics can improve the bacterial diversity in the mouth and in the GI-tract.

6) Probiotics can improve the digestion, by for example, increasing the absorption of iron and other essential trace elements from the GI-tract.

7) Probiotics can contribute to the break-down of indigestible carbohydrates to short-chain-fatty-acids with physiologically beneficial effects.

8) Probiotics can contribute to the break-down of indigestible polyphenols into digestible compounds with beneficial anti-oxidative, and antimicrobial properties.

Probiotics are living organisms and as such they can exercise several fundamentally different actions on different target-functions in the body. All probiotic strains are not able to fulfil all sorts of missions. The capacity of different strains for acting on the different target-function can vary within wide limits.

A rational for combining different probiotic strains in a mixture could be that they are differently good in affecting different functions. For example, one strain can be especially good in modulating the immune system by targeting dendritic cells and T-cells, while another can have a strong influence on the epithelial barrier function of the mucosa.

Probiotics have been asserted to mitigate a variety of dysfunctions or diseases, but the reliability of the scientific evidence of the different claims can vary extensively.

There is reasonably reliable evidence for the ability of probiotics to counteract, at least to some extent, the following dysfunctions or diseases:

— diarrhoea due to dysbiosis (unbalanced microbiota),
— rota virus diarrhoea,
— common cold,
— functional bowel disorders,
— inflammatory bowel diseases,
— liver injuries,
— leaky gut and systemic inflammation,
— metabolic syndrome,
— autoimmunity,
— oestoperosis
— iron deficiency.

Immune modulation

Probiotics can affect the immune system which can be visualised by changes in the levels of different immunological agents in blood and tissue, and these agents can be used as markers for the status of the immune system. Immunological markers are frequently followed in all sorts of clinical trials, and in animal models.

Below are two examples of clinical trials with probiotics where the modulation of the immune system has been followed.

In the study about eczema, the immune markers have been followed as a help to explain the studied end-point which is a lower rate of the disease.

Study-example: incidence of eczema in weaning infants

In a double-blind, placebo-controlled randomized intervention trial, 90 infants, from 4 to 13 months of age, were fed cereals with *Lactobacillus* F19, and 89 were fed cereals without probiotics. The incidence of eczema and Th1/Th2 balance were recorded at 13 months of age.

The ratio of interferon-gamma (IFN-gamma) to interleukin 4 (IL4) mRNA expression levels in polyclonally stimulated peripheral blood T cells was used as a proxy for immune balance. Total and specific IgE serum levels were also assessed.

The cumulative incidence of eczema was 11% and 22% in the probiotic and placebo groups, respectively ($p<0.05$). The IFN-gamma/IL4 mRNA ratio was higher in the probiotic group compared with the placebo group ($p<0.05$), without any changes in IgE concentrations.

Thus, feeding *Lactobacillus* F19 during weaning could be an effective tool in the prevention of early manifestation of eczema. The higher Th1/Th2 ratio in the probiotic group, compared with the placebo group suggests enhancing effects of *Lactobacillus* F19 on the T cell-mediated immune response.

[West *et al.* (2009). Probiotics during weaning reduce the incidence of eczema. *Pediatric allergy and immunolog* 20:430—437]

Note:
"*Lactobacillus* F19" should be written *Lacticaseibacillus paracasei* F19.

T-helper cells (Th) are a sub-group of lymphocytes involved in establishing and maximizing the capabilities of the immune system. Th-cells have no cytotoxic or phagocytic activity. Proliferating Th-cells differentiate into two major subtypes, Th1 and Th2. The different types produce different cytokines, and affect different lymphocytes.

In contrast, the study about immune response under stress (below) only demonstrate effects on immunological markers without link to any particular dysfunction or disease. Among several measured blood parameters, significant changes were only seen for the number of lymphocytes and CD56. The practical relevance for the health status by this change in immune parameters remains unclear.

The immune system is highly complex with a vast number of different markers that often are related to each other, and a comprehensive and full knowledge of function and interplay is lacking, or at least highly muddled.

Study-example: immune response under stress

The effect of milk fermented with yogurt cultures plus *Lactobacillus casei* DN-114001 on the immune system of subjects under academic examination stress was evaluated in a randomized, controlled and parallel study.

University students divided into two groups, received during 3 weeks prior to, as well as the 3-week duration of the examination period. The control group (63 students) consumed 200 ml of semi-skimmed milk each day, and the treatment group (73 students) consumed two 100 ml portions per day of fermented milk. Anxiety and immunological measurements were monitored at baseline and at end of study (after 6 weeks).

The results were expressed as the differences between the data obtained at the start and the end of the study. There was no significant treatment effect in anxiety.

There was a significant treatment effect ($P<0.05$) on the mean change in absolute number of lymphocytes, which decreased in the control group and increased in the treatment group.

There was a significant effect ($P<0.05$) on the change in absolute numbers of CD56 cells.

Thus, milk fermented with yogurt cultures plus *Lactobacillus casei* DN-114001 was able to modulate the number of lymphocytes and CD56 cells in subjects under academic examination stress.

[Marcos *et al.* (2004). Milk fermented with yoghurt cultures plus *L. casei* DN-114001 on the immune response of subjects under academic examination stress. *Eur. J. Nutr.* 43:381-389]

Note:
The tested fermented milk was Actimel (Danone), and contained *Lactobacillus delbrueckii* subspecies *bulgaricus* (10^7 CFU/ml), *Streptococcus thermophilus* (10^8 CFU/ ml) and *Lacticaseibacillus paracasei* DN-114001 (10^8 CFU/ml).

CD stands for "clusters of differentiation".
CD56 positive cell means a T-cell with the membrane protein CD56 on the surface. CD56 is the same as NCAM (Neural Cell Adhesion Molecule) which is a binding glycoprotein expressed on the surface of neurons, glia, skeletal muscle and natural killer cells. NCAM has been implicated in cell-to-cell adhesion, neurite outgrowth, synaptic plasticity, and learning and memory.

Changes in anxiety, in the above study would have been a nice end-point, but no significant effects in anxiety were reached.

Immunological markers are insufficient end-points for clinical trials. However, changes in immunological markers are helpful when explaining causes and mechanisms.

The immunological markers should be combined with endpoints revealing undisputable health improvements.

Some immunological terms

Immune-competent cells are macrophages, dendritic cells, B-lymphocytes, and T-lymphocytes. They are found in Peyer's patches and other gut-associated lymphoid tissue.

Parenchymal cells (hepatocytes) and the non-parenchymal cells, that is, endothelial cells, Kupffer cells, hepatic stellate cells and Pit cells (Pit cells, are liver-specific natural killer cells).

Monocytes are big white blood cells in the blood stream. Monocytes are produced by stem-cells in the bone marrow.

Macrophages are white blood cells within tissue. Macrophages are converted monocytes, and they perform phagocytosis.

Dendritic cells are also converted monocytes. They are messengers between the innate and adaptive immune systems. Dendritic cells present antigens.

There are three major types of lymphocytes, T-cells, B cells and NK cells.

T-cells are lymphocytes where the binding capacity to body-components has been evaluated by a selection process in thymus.

B cells are lymphocytes that produce antibodies, present antigens and can develop into memory cells. They are essential component of the adaptive immune system.

NK cells are cytotoxic lymphocytes. They are large granular lymphocytes, and are a major component of the innate immune system.

Neutrophil granulocytes (neutrophil or polymorphonuclear neutrophil = PMN) are the most abundant type of white blood cells and form an essential part of the innate immune system. Neutrophils are motile, and phagocytic. They express and release cytokines, and have a key-role in the defence against invading pathogens. Neutrophils can disarm and kill many types of microbes.

Eosinophil granulocytes (eosinophils) are white blood cells that produce granule proteins, reactive oxygen species, lipid mediators, enzymes, growth factors and cytokines. They are active against multicellular parasites and virus, and control mechanisms associated with allergy and asthma.

Basophil granulocytes (basophils) are white blood cells that are susceptible to staining by base dyes and contain large cytoplasmic granules. They appear in many inflammatory reactions, particularly in allergy. Basophils contain heparin (prevents blood from clotting) and histamine (promotes blood flow to tissues).

Mast cells (mastocytes) are immune cells resident in several types of tissue and contain many granules rich in histamine and heparin. They are intimately involved in wound healing and the defence against pathogens. Mast cells play a role in allergy and anaphylaxis (hypersensitivity).

NF-κB stands for "nuclear factor kappa-light-chain-enhancer of activated B cells". NF-κB is a protein complex that controls the transcription of DNA.
NF-κB plays a key role in regulating the immune response to infections.
Incorrect regulation of NF-κB is linked to cancer, inflammatory and autoimmune diseases, septic shock, viral infection, and improper immune development.

Toll-like receptors (TLRs) are expressed on the surface of macrophages, dendritic cells, lymphocytes and epithelial cell surface. Toll-like receptors (TRLs) are so named because of their similarity to a receptor first identified in the fruit fly *Drosophila melanogaster* by German scientists, a protein coded by the Toll-gene ("toll" means fantastic in German). At least ten types of human TLRs are known. TLRs are expressed in most tissues in healthy adult individuals.

Diarrhoea due to dysbiosis

Antibiotics can disturb the bacterial balance of the gut which leads to **antibiotic associated diarrhoea**. Antibiotic associated diarrhoea, also called pseudomembranous colitis is an infection in the colon caused by an unfavourable change in the gut-microbiota due to the antibiotic treatment.

The anaerobic spore-former *Clostridium difficile* (CN: *Clostridioides difficile*) is the major cause of antibiotic associated diarrhoea, that is, *C. difficile* colitis. The incidence of antibiotic associated diarrhoea in hospitalized patients can be expected to be somewhere between 3—29 %, and *C. difficile* accounts for 20—60 % of these diarrhoeas.

C. difficile produces several virulence factors, that is, enterotoxin, cytotoxin and an inhibitor for bowel motility.

Clinical symptoms range from mild diarrhoea to toxic megacolon which in turn may lead to bowel perforation of the intestinal wall.

C. difficile associated diarrhoea is usually treated with the antibiotics (metronidazole or vancomycin).

Treatment with probiotics can to a certain extent counteract antibiotic associated diarrhoea and *C. difficile* associated diarrhoea. The preventing effect of probiotics is more reliable than the effect on an already established diarrhoea.

The efficiency of treatment is highly dependent on:
(*i*) which probiotic strain that is used, and on
(*ii*) the use of a sufficient dose of metabolic active probiotic cells.

Different strains of, for example, *L. paracasei*, *L. rhamnosus* and *L. plantarum* have been shown to have effect in randomized placebo-controlled studies.

Study-example: Antibiotic associated diarrhoea

In a randomized double-blind placebo-controlled trial, 100 g active product, or a placebo product, were given twice per day during antibiotic treatment of 135 hospital patients, and one week after termination of the antibiotic treatment. The active product contained 10^8 CFU/ml "*L. casei*" (*L. paracasei*) DN-114001, 10^8 CFU/ml *Streptococcus thermophiles*, and 10^7 CFU/ml *L. delbrueckii* subsp. *bulgaricus*.

Only seven out of 57 (12%) of the patients in the probiotic group developed diarrhoea, compared with 19 out of 56 (34%) in the placebo group. No one in the probiotic group, and 9 out of 53 (17%) got *C. difficile* associated diarrhoea in the placebo group.

The conclusion was that the used probiotic mixture had a preventive effect, and reduced the incidence of antibiotic associated diarrhoea, and *C. difficile* associated diarrhoea.

[Hickson *et al*. (2007). Use of probiotic *Lactobacillus* preparation to prevent diarrhoea associated with antibiotics: randomised double blind placebo controlled trial. *British Medical Journal* 335:80—83]

Virus

Rotavirus diarrhoea

Probiotics can have effects against viral diarrhoea, which primarily have been shown for rotavirus gastroenteritis. Thus, probiotics do not only contradict adverse effects of bacteria in the large bowel, but also that of viruses in the small bowel. Rotavirus is one of several viruses that can cause the type of infections that in every day talk are referred to as "stomach flu".

The name rotavirus refers to the wheel-like appearance of the virus that has a double-stranded RNA-genome. Rotavirus is the major cause of viral gastroenteritis in infants and young children.

The infection is spread by the faecal-oral route, and rotavirus infects the epithelial cells lining in the small intestine with an incubation period of 1—2 day.

The duration of the disease is 4—7 day, and the symptoms are vomiting, diarrhoea, fever and stomach-pain. In small children rotavirus gastroenteritis can lead to severe dehydration.

There have been a multitude of studies given evidence that probiotics can mitigate the symptoms and reduces the duration of rotavirus gastroenteritis. Strains of different lactibacilli and bifidobacteria have been shown to have effects, but most thoroughly studied is the strain *Lacticaseibacillus rhamnosus* GG.

Probiotics can activate the unspecific immunological defence during the rota virus infection, and increases the level of specific antibodies in the following-up period. Thus, the probiotic effect against rotavirus is presumably an effect mediated mainly by the immune system.

However, the actual cause of the probiotic effects, seen on the outcome of the disease, may also involve the status of the mucosa. A mucosa in good condition will resist the viral attack more successfully than a mucosa in poor condition.

WINTER VOMITING DISEASE

Calicivirus gastroenteritis is caused by RNA virus of the viral genus *Norovirus* and the viral family *Caliciviridae*. Calicivirus is preferably transmitted in institutional settings and by contaminated air, food and water.

Clinical features of calicivirus gastroenteritis are diarrhoea and vomiting, generally without fever. The incubation period is 2—3 d, with a duration of 1—11 d.

In perspective of the effects of probiotics against rotavirus gastroenteritis, a relevant question is: Do probiotics also help against winter vomiting disease? Unfortunately, no conclusive answer is available.

Common cold

Common cold, or just cold, is a viral upper respiratory tract infection (also called acute viral rhinopharyngitis or acute coryza) that can be caused by several different viruses.
Examples of different viruses giving rise to common cold are:
rhinoviruses (in about 30–50 % of cases),
coronaviruses (10–15 %) and
influenza virus (5–15 %),
but common cold can also be caused by other viruses as human parainfluenza viruses, human respiratory syncytial virus, adenoviruses, enteroviruses or metapneumovirus.

Cold is the most frequently occurring, human infectious disease. Adults have on average 2—4 infections per year and children have 6—12 infections per year. Due to the fact that there are so many different types of viruses and that the viruses tend to mutate, it is impossible to become immune to all colds.

Without specifying which viruses that have been active and responsible for the infection, consumption of probiotics has in several, good randomized studies showed effects against common cold in both children and adults.

Probiotics can reduce the severity of symptoms and duration of illness. However, the majority of the studies failed to affect the incidence of the disease.

Strains of both bifidobacteria and lactobacilli have been used successfully as probiotics against common cold.

The traditional view on probiotics is that it is supposed to have beneficial effects in the gut, and maybe it is a bit unexpected to see effects in the upper respiratory tract. However, the common cold viruses infect the mucous membranes, and the general action of the probiotics on exposed mucous membranes ought to be similar regardless of the location of the membranes. Probiotics administrated orally will for certain reach the mouth and throat, and presumably also to some extent also reach the mucous membranes of the nose. And maybe, some even will reach the lungs, mediated by aerosols in the air of inhalation.

The cause of the probiotic effect on respiratory virus infections is not clarified, but it seems reasonable to assume that improving effects:
(*i*) on the immune system, and
(*ii*) on the condition of the mucosa
can be involved.

Study-example: Common cold

In a randomized double blind and placebo-controlled trial, 310 volunteers with a history of two or more episodes of common cold during the last six months, receive either a daily dose of 10^9 CFU of a freeze-dried blend of *L. plantarum* HEAL9 (=DSM 15312) and *L. paracasei* 8700:2 (=DSM 13434), or a placebo, over a period of 12 weeks. Occurrence and severity of cold episodes was individually reported in a diary, scoring 13 different symptoms of common cold over a period of 7 days after onset of the cold. The number of episodes was similar in both groups, but cold episodes in the active group were significantly shorter than in the placebo group. Also, the total sum of scored symptoms was significantly lower in the active group as compared with the placebo group.
In conclusion, a daily intake of the probiotic supplement for 12 weeks efficiently alleviated symptoms of common cold and the duration of cold episodes.

[Busch *et al.* (2013). Randomized, double blind and placebo controlled study using a combination of two probiotic lactobacilli to alleviate symptoms and frequency of common cold. *Food and Nutrition Sciences* 4: 13-20 http://dx.doi.org/10.4236/fns.2013.411A003]

Functional bowel disorders

DYSFUNCTIONS

Examples of common, transient symptoms of intestinal disorders are bloating and flatulence, abdominal pain, diarrhoea and constipation. If these conditions have occurred for at least 6 months, and if they are present for more than 3 days per month during a period of 3 months, the medical diagnosis is *Functional Bowel Disorder* (FBD).

FBD is then subdivided into *Irritable Bowel Syndrome* (IBS), functional bloating, functional constipation, and functional diarrhoea. Different gut associated conditions are solely diagnosed by their symptoms and may be caused by different reasons, or according to medical terminology, they can have multiple aetiologies.

The criteria of FBD and IBS are consensus criteria that gradually have been worked out by the Rome Foundation which is an independent organization that provides support for activities designed to create scientific data and educational information to assist in the diagnosis and treatment of functional gastrointestinal disorders.

Consumers with Functional bowel disorder (FBD) is an interesting target-group for the probiotic business, due to the fact that FBD is a frequently occurring problem in urban populations, and a problem that cause major discomfort for otherwise healthy individuals. Thus, if a company want to make a general claim that a certain probiotic supplement or food-product is good for the gut health, this claim can be proved by backing it up with studies on FBD.

FBD is a diagnosis of discomfort and, formally, not a disease. This makes it possible to use data from FBD-studies as argument in connection to health promoting foods without risking to be accused by the authorities to market a pharmaceutical drug as food.

Pharmaceuticals on the other hand, must be registered (and evaluated) as such and is then not allowed in the food segment.

An advantage with claims directed towards FBD is that the consumer suffering from FBD can get a direct self-evaluation of the efficiency of the product on hers or his FBD-symptoms by just eating the product and experience the result. A positively outcome will presumably lead to continues purchase of the product.

A problem to get scientific backup for claims about FBD related symptoms is the lack of knowledge about the reasons for the symptoms, and the linked difficulties to select research persons with uniformed FBD-characteristics.

IRRITABLE BOWEL SYNDROME (IBS)

Irritable bowel syndrome (IBS) is a functional bowel disorder (FBD) in which abdominal pain or discomfort is associated with defecation or a change in bowel habit, and with features of disordered defecation.

IBS is a commonly occurring syndrome of disorders of unknown cause. The absence of strict pathogenic features has for long made IBS to an illness without a proper diagnosis.

IBS is a chronic relapsing condition that perhaps occurs in most adults at some point in their lives. Symptoms begin before the age of 35 in 50% of patients.

Throughout the world, about 10%–20% of adults and adolescents have symptoms consistent with IBS. In most countries more females than men suffer from IBS.

A considerably numbers of clinical trials have been performed with the aim to evaluate the effect of probiotics in subjects with IBS or IBS-related problems. Some studies have shown clearly significant effects while others have failed to show any effect. The outcome of the studies is dependant of what type of probiotics that has been administrated, and the uniformity of the tested group of patients.

Successful outcomes have been reached with both single-strain probiotics and multiple-strains probiotics. The applied strains in the single-strain-probiotics are mostly carefully characterized with a clear identity, while the multi-strain-probiotics often include strains of more unclear identity.

Two well-known probiotic strains that separately have showed beneficial effects in randomized, double-blind placebo-controlled studies on IBS are *Bifidobacterium animalis* DN-173010 (Danone) and *Lactoplantbacillus plantarum* 299v (=DSM 9843; Probi AB, Sweden), respectively.

Study example 1: IBS

In a randomized, double-blind, controlled trial, the effect of a fermented milk product containing *B. animalis* DN-173010 on the health-related quality of life and on symptoms of irritable bowel syndrome (IBS) was evaluated in adult patients in primary care. 274 subjects with constipation-predominant IBS consumed 125 g product (Activia, Danone) per day, or placebo for 6 weeks. The probiotic dose was 10^{10} CFU of *B. animalis*, plus the yoghurt starter, that is *S. thermophilus* and *L. bulgaricus* (10^9 CFU).

The results showed a beneficial effect on "health-related quality of life" discomfort score, and an increased stool frequency in subjects with less than 3 stools per week at the start of the treatment.

[Guyonnet *et al.* (2007). Effect of a fermented milk containing *Bifidobacterium animalis* DN-173010 on the health-related quality of life and symptoms in irritable bowel syndrome in adults in primary care: a multicentre, randomized, double-blind, controlled trial. *Alimentary Pharmacology & Therapeutics* 26:475-486]

Study example 2: IBS

In a double blind, placebo-controlled, parallel-designed study, 214 IBS-patients were randomized to daily receive either one capsule of freeze-dried *L. plantarum* 299v (=DSM 9843) or placebo for 4 weeks. Frequency and intensity of abdominal pain, bloating and feeling of incomplete rectal emptying were assessed weekly on a visual analogue scale, while stool frequency was calculated. After 4 weeks there was a significant decrease in pain severity and daily frequency in the active group compared with the placebo. Similar results were obtained for bloating. Furthermore, 78% of the patients in the active group scored the symptomatic effect as excellent or good, compared with 8 % in the placebo ($P<0.01$).

The conclusion of the authors is that a four-week treatment with *L. plantarum* 299v provided effective relief of symptoms in IBS patients, especially of abdominal pain and bloating.

[Ducrotté *et al.* (2012). Clinical trial: *Lactobacillus plantarum* 299v (DSM 9843) improves symptoms of irritable bowel syndrome. *World Journal of Gastroenterology* 18:4012-4018]

Inflammatory bowel disease

CHRONIC INFLAMMATION

Inflammatory bowel disease (IBD) is a chronic inflammation somewhere along the gastro-intestinal (GI) tract. It can be limited to the large bowel (ulcerative colitis) or it can be situated anywhere along the GI-tract (Crohn's disease).

Ulcerative colitis is a relatively superficial ulcerative inflammation, while Crohn's disease is a transmural, granulomatous inflammation. There are also several other forms of IBD, but they are occurring less frequently, and they are not always so well defined.

IBD is thought to be due to an abnormal and uncontrolled immune response to normally occurring constitutes of the intestine. The cause (aetiology) of IBD is unknown.

Microbial agents appear to be involved in the pathogenesis, and intestinal bacteria seem to be an important factor in the development and chronicity. There is a complex interaction between bacteria, mucosa and immune system but this interaction is far from clear.

Probiotics might have a potential for mitigate inflammation and symptoms in IBD patients.

CROHN'S DISEASE

Crohn's disease is regarded as an autoimmune disease. It usually hits at 15—25 years of age, with a prevalence of 30—50 individuals per 100 000. Crohn's disease can appear anywhere in the digestive tract, from the mouth to the anus. Though, it often starts in terminal ileum.

The cause of Crohn's disease is unknown. The idea is that it is a combination of environmental factors and genetic predisposition that coincides, and results in a malfunction in the innate immune systems where chronic inflammation being caused by adaptive immunity that tries to compensate for the reduced function of the innate immune system.

There is no known cure for Crohn's disease. Treatment is restricted to controlling symptoms, maintaining remission and preventing relapse, and here can perhaps probiotics be of some help by counteract pro-inflammatory bacteria in the digestive tract.

It is clear that the microbiota of the digestive tract can aggravate the intestinal inflammation, especially certain aggressive components of the microbiota as *Escherichia coli* and *Bacteroides fragilis*.

Lacticaseibacillus rhamnosus GG and *Lactobacillus. johnsonii* LA1 have both been tested on patients with Crohn's disease, but with discouraging results.

Study example: Crohn's disease

Intestinal epithelial cells were taken from patients with Crohn's disease, and were in the laboratory exposed to either *Escherichia coli* (strain ATCC 35345) or *Lacticaseibacillus casei* (strain DN-114001); after which the immunological response of the cells was studied. An exposure to *L. casei* decreased the secretion of TNF-alpha, IFN-gamma, IL-2, IL-6, IL-8, and CXCL1, and down regulated expression of IL-8, IL-6, and CXCL1 in the human tissue while, in contrast, an exposure to *E. coli* up-regulated the expression of all the cytokines.

Furthermore, live *L. casei* counteracted the pro-inflammatory effects of *E. coli* on inflamed mucosa by specific down-regulation of key pro-inflammatory mediators.

[Llopis *et al.* (2009). *Lactobacillus casei* downregulates commensals' inflammatory signals in Crohn's disease mucosa. *Inflammatory Bowel Diseases* 15:275-282]

Pro-inflammatory cytokines (cytokine = interleukin, IL)

Tumour necrosis factor (TNF) alpha is produced by monocytes/macrophages and activated T-cells.

Interferon (IFN) gamma is a key cytokine in chronic inflammation (originally, proteins that increase the resistance of body cells against viruses).

IL-2 stimulates the growth of T-cells (and is also produced by T-cells).

IL-6 initiates production of acute phase proteins and takes part in finish-off the acute inflammation (produced by macrophages and T-cells).

IL-8 activates white blood cells to pass out from the blood vessel (produced by macrophages and epithelial cells after stimulation of bacteria or bacterial products).

CXC-chemokines binds to neutrophlic granulocytes and have similar effects as IL-8.

ULCERATIVE COLITIS

Ulcerative colitis causes ulcers, or open sores in colon, and gives diarrhoea mixed with blood. The disease starts from below and moves upwards in colon, and increases the risk for colorectal cancer.

The cause of ulcerative colitis is unknown, but the hypothesis is that it depends on genetic susceptibility that is triggered by environmental factors. Ulcerative colitis is sometimes treated as an autoimmune disease, but there is no consensus about ulcerative colitis being an autoimmune disease.

Ulcerative colitis is characterized by periods of remission marked by episodes of clinical relapse caused by acute colonic and/or rectal inflammation. Treatment is primarily aimed at reducing inflammation during relapse and at prolonging the time spent in remission of clinical symptoms.

During the acute phase of inflammation, macrophages, neutrophils and eosinophils are infiltrating the layer of connective tissue which separates the innermost layer of epithelial cells from a layer of smooth muscle tissue; this layer is called the lamina propria.

Aggregating neutrophils lead to the formation of abscesses. Activated dendritic cells and macrophages secrete cytokines that trigger and differentiate T cells, and activating the adaptive immune response.

Patients with ulcerative colitis seem to have a higher number of bacteria associated to the mucosa than healthy persons. The intestinal microbiota in patients with active ulcerative colitis can be expected to be less diverse than in healthy subjects, and the proportion of inflammation inducing bacteria can be expected to be higher. The abundance tends to increase for taxa as *Enterobacteriaceae* (for example, *E. coli* and *Enterobacter*), *Bacteroides fragilis*, and *Pseudomonas aeruginosa.*

Hydrogen sulphide has also been implicated in the pathogenesis of ulcerative colitis, and sulphate-reducing bacteria have received attention due to their ability to reduce sulphate to sulphide as a by-product of their respiration. Hydrogen sulphide is cell toxic and freely permeable to cell membranes and it also inhibits butyrate oxidation in the colonic epithelial cells (colonocytes).

Probiotics has in intervention studies been given to patients with mild to moderate ulcerative colitis, and has in active phase of the disease decreased the clinical activity index, induced remission and prevented relapses.

Administered probiotics have in these studies often been mixtures of strains of different genus and species.

It should be stressed that the evidence for probiotic therapy against ulcerative colitis is so far weak.

Liver injuries

Fibrosis and cirrhosis

The gut and the liver are closely connected, and a well-functioning link between the two is dependent on both an intact intestine and a healthy liver in balance with respect to immunological response and metabolism of endogenous and exogenous compounds.

The liver is an important site for bacterial phagocytosis and clearance. It contains the largest population of tissue-bound macrophages of the body. The macrophages of the liver are called Kupffer cells.

Chronic liver injury is associated with the development of fibrosis, since repeated and continuous cellular damage of liver cells leads to the activation of hepatic stellate cells (star shaped nerve cells which are involved in liver fibrosis). The hepatic stellate cells produce extracellular matrix proteins which build up scar tissue in the liver.

An advanced stage of hepatic fibrosis is cirrhosis, in which functional liver tissue is largely replaced by scar tissue and regenerating nodules. Cirrhosis is mostly caused by alcoholism, hepatitis B and C, and fatty liver disease.

Fibrosis can promote bacterial overgrowth in the small intestine, weakened barrier function in the gastro-intestinal tract, due to restricted intestinal motility, decreased bile acid production, and decreased nutrient absorption and availability.

Disturbances of the gut-microbiota are prevalent in patients with chronic liver disease. Patients with chronic liver disease can be expected to suffer from increased bacterial translocation, in combination with failure of immune defence mechanisms, and predisposition to bacterial infections.

The increased translocation of intestinal bacteria leads to bacterial invasion of the mesenteric lymph nodes, and endotoxins (from gram-negative bacteria) in the portal blood, and if the liver becomes overrun, bacteria or parts of bacteria as endotoxins will go out in the systemic circulation (the blood). Thus, an impaired gut barrier function will have consequences for the liver status, and will increase the blood pressure in the portal vein, and contribute to the progression of chronic liver damage.

Generally, an adverse gut microbiota seems to be able to induce liver disease by means of increased portal delivery of lipopolysaccharides (LPS) and other endotoxins, which leads to activation of the stationary macrophages in the liver (Kuppfer cells), and induction of production of TNF-*beta* (similar to TNF-*alpha*), and subsequent activation of hepatic stellate cells.

The ability of probiotics to improve the intestinal microbiota, stimulate the immunological system in a favourable direction and to improve the barrier function of the mucosa, imposes effects that can mitigate the harmful effects of liver disease. But probiotics can never by its own account cure a liver disease. However, long term probiotic consumption is probably a preventing factor when it comes to liver failure.

MINIMAL HEPATIC ENCEPHALOPATHY

Minimal hepatic encephalopathy (MHE) is a disorder in patients with cirrhosis, and a disorder that seriously can impair daily functioning and quality of life. Increased level of ammonia in the blood is most probably a key factor in the pathogenesis.

Probiotics can decrease the concentration of ammonia in the blood, and decrease the endotoxin levels in patients with liver cirrhosis.

ALCOHOL-RELATED LIVER DISEASE

Chronic ethanol consumption causes changes in the liver, as fatty liver, inflammation and cirrhosis.

Inflammation associated to accumulated fat in the liver is called steatohepatitis, and is characterised by infiltration of monocytes, macrophages, neutrophils, and lymphocytes; they occur as a consequence of activation of inflammatory mediators.

During alcoholic steatohepatitis, TNF-α, IL-6, and IL-8 levels in serum increase and the increase correlates with markers for the acute-phase response, liver function, and clinical outcome.

Alcoholic steatohepatitis leads also to an increase in the load of gut-bacteria, and to a change in the composition of the microbiota; a change that may results in overgrowth of the small intestine, in endotoxaemia (endotoxins in the blood), and in increased bacterial production of acetaldehyde from ethanol.

Acetaldehyde is more toxic than ethanol, and an increased concentration of acetaldehyde will accentuate the liver damage. Moreover, ethanol and acetaldehyde affect intracellular signal-transduction pathways leading to the disruption of epithelial tight junctions, which increases the paracellular permeability of macromolecules. Thus, an impaired gut barrier function is a contributing factor in the progression of chronic liver damage.

Probiotics can have beneficial effects on patients with alcohol-related liver disease, and strains of *Bifidobacterium bifidum*, *Lactiplantibacillus plantarum* and *Lacticaseibacillus paracasei* have been used with promising results.

Study example: alcohol-related liver disease

Patients with mild alcoholic induced hepatitis, and with the diagnosis alcoholic psychosis was in an open-label, prospective, randomized, clinical trial receiving *Bifidobacterium bifidum* (no strain label is given) and *L. plantarum* 8PA3 for 5 days (32 patients), or standard therapy for abstinence plus vitamins (34 patients). The average daily volume of alcohol consumed 1—2 weeks before the recruitment was around 750 ml per day of Russian vodka (40% ethanol).

Analyses were performed before treatment (baseline) and after treatment. Measurements of liver enzymes were performed at baseline and again after termination of therapy.

The probiotic treatment resulted in reduction of the levels of alanine aminotransferase (ALT), aspartate aminotransferase (AST), gamma glutamyl transpeptidase, lactate dehydrogenase, and total bilirubin in the blood.

[Kirpich *et al.* (2008). Probiotics restore bowel flora and improve liver enzymes in human alcohol-induced liver injury: a pilot study. *Alcohol* 42: 675–682]

Indicators of liver failure

Alanine aminotransferase (ALT) is an indicator for the damage of liver cells (hepatocytes).

Aspartate aminotransferase (AST) is an indicator for the damage of liver cells, but AST can also be found in the heart and different muscle tissues.

Bilirubin is a product from destroyed red blood cells that will increase if the liver is failing, and is a measure of hepatic transport function and the severity of jaundice (icterus).

Gamma-glutamyl transpeptidase (GGTP) catalyses the removal of glutamyl groups of aromatic amines to other amine receptors on peptides or amino acids; most GGTP is found in the kidney, with smaller amounts in the liver and heart.

Lactate dehydrogenase (LDH) is an enzyme found in nearly all living cells, and catalyses the conversion of lactate to pyruvate and back, as it converts NAD+ to NADH and back.

Non-Alcoholic Fatty Liver Disease

Accumulation of fat and liver inflammation

Non-alcoholic fatty liver disease (NAFLD), or as it also is called Metabolic associated fatty liver disease (MAFLD), is an inflammation in the liver due to accumulation of fat, when the fat accumulation is not due to an excessive alcohol consumption.

NAFLD is a comprehensive term for a chain of conditions with successively increasing the severity. The different stages in liver injury are:
steatosis (fat accumulation),
steatohepatitis (inflammation with concurrent fat accumulation; called non-alcoholic steatohepatitis [NASH]),
liver fibrosis (excessive accumulation of extracellular matrix proteins including collagen),
cirrhosis (failing liver function, and hepatocellular carcinoma [carcinoma stands for any malignant cancer that arises from epithelial cell]).

In the initial phase of NAFLD, the healthy liver accumulates fat (steatosis), mainly as a consequence of peripheral resistance to insulin. The insulin-resistance increases the transport of fatty acids from adipose tissue to the liver.

Steatosis renders hepatocytes susceptible to further obstacles. Once steatosis is established there are several factors that can lead to an inflammatory process that can result in hepatocellular degeneration and fibrosis.

Negative factors for the establishment of steatosis, besides ethanol are:
(*i*) lipopolysaccharides (LPS) leaking out from the intestinal microbiota to the liver,
(*ii*) oxidative stress,
(*iii*) cytokines aggravated dysfunction of liver cells.

NAFLD is related to insulin resistance and the metabolic syndrome, and strongly linked to obesity and type 2 diabetes. It is also associated with hyperlipidaemia (too much fat in the blood), and coeliac disease.

NAFLD can also be caused by:
(*i*) different types of medications,
(*ii*) exposure to environmental toxins,
(*iii*) total parenteral nutrition,
(*iv*) surgical interventions in the digestive tract.

Microbiota

An endogenous factor that can contribute to the pathogenesis of NAFLD is the microbiota of the digestive tract. For example, by producing ethanol which enhance the oxidative stress on the liver.

Gram-negative bacteria in the gut may also affect hepatic oxidative stress through release of LPS, leading to production of inflammatory cytokines by stimulation of luminal epithelial cells and Kupffer cells. Kupffer cells are the main producers of TNF-α, a central mediator in the pathogenesis (development) of NASH.

Probiotics

It can be speculated in what way probiotics can counteract the development of NAFLD. One suggestion is that the probiotic bacteria replace aggravating bacteria in the digestive tract, which in turn can decrease the production of proinflammatory cytokines, like TNF-α.

An alternative explanation could be that the probiotic bacteria improve the epithelial barrier function and thereby decreases the exposure of the liver to LPS and to ethanol.

However, despite the rationale for the possible therapeutic role of probiotics, and lots of promising results from different types of animal models, convincing results from placebo-controlled trials on humans are scant.

Study example: NAFLD

The therapeutic effect of a mixt-strain probiotics on different liver diseases, including a small group of NAFLD patients, was performed. The probiotic mixture is claimed to include *Lacticaseibacillus casei*, *Lactiplantibacillus plantarum*, *Lactobacillus acidophilus*, *Lactobacillus delbrueckii* subsp. *bulgaricus*, *Bifidobacterium longum*, *Bifidobacterium brevis*, *Bifidobacterium infantis* and *Streptococcus thermophiles*, and was given to patients with non-alcoholic fatty liver disease (NAFLD; 22 patients); alcoholic liver cirrhosis (AC; 20 patients), and hepatitis C virus infection (HCV; 36 patients) in a dose of totally $5x10^{11}$ CFU per day for 3 months.

Different markers of the liver function, oxidative stress and inflammation were measured before the start of the probiotic treatment and after 3 months of treatment.

Aspartate aminotransferase (AST), alanine aminotransferase (ALT) and bilirubin decreased in all patient categories. In the NAFLD and the AC patients the oxidative stress decreased (seen as decreased concentration of malondialdehyde), and in the AC patients also the inflammation decreased (seen as decreased levels of TNF-alfa, IL-6 and IL-10)

[Loguercio, *et al.* (2005). Beneficial effects of a probiotic VSL#3 on parameters of liver dysfunction in chronic liver diseases. *Journal of Clinical Gastroenterology* 39:540–543]

Malondialdehyde (MDA) is a marker for oxidative stress.

Reactive oxygen species degrade poly-unsaturated lipids, forming MDA, which cause toxic stress in cells and form covalent protein adducts.

Patients in intensive care

SYSTEMIC INFLAMMATION

Intensive care medicine, or critical care medicine, is a discipline dealing with the support of life maintenance and organ support in patients who are critically ill. It is obvious that there are many factors to watch out for in these situations, but one factor of relevance for probiotics is the problem with a leaky gut leading to an excessive inflammation that can put the patient in danger.

Probiotics can to some degree decrease translocation of bacteria and bacterial components as LPS from the gut, and thereby decrease systemic inflammation.

Study example: Systemic inflammation

A decrease in systemic inflammation by administration of *Lactplantibacillus plantarum* 299v has been seen in critically ill patients in the form of a decreased level of IL-6 in the blood. 103 critically ill patients were randomised to receive *L. plantarum* 299v in the fruit beverage ProViva, in addition to conventional therapy. The median value of the daily volume of ProViva was 213 ml, corresponding to a dose of about 10^{10} CFU of *L. plantarum* 299v per day. The median for the duration of intake was 9 days.
Enteral administration of *L. plantarum* 299v to critically ill patients resulted in a late attenuation of the systemic inflammatory response.

[McNaught, *et al.* (2005). A prospective randomised trial of probiotics in critically ill patients. *Clinical Nutrition* 24:211—219]

REDUCED COLONISATION OF *C. DIFFICILE*

The incidence of *Clostridium difficile*- (CN: *Clostridioides difficile*) -associated disease (CDAD) in hospitalised patients is increasing. Critically ill patients are often treated with antibiotics and are at risk of developing CDAD.

Lactiplanibacillus plantarum 299v can reduce recurrence of CDAD in patients treated with antibiotics in the intensive care.

Study example: Reduced colonisation of *C. difficile*

The aim was to investigate patients in an intensive care unit (ICU) with respect to the impact of *Lactobacillus plantarum* 299v on *C. difficile* colonisation and on gut permeability and parameters of inflammation and infection.
Twenty-two ICU patients were given a fermented oatmeal gruel containing *L. plantarum* 299v, and 22 received an equivalent product without the bacteria. Faecal samples for analyses of *C. difficile* and *L. pantarum* 299v were taken at inclusion and then twice a week during the ICU stay.
Colonisation with *C. difficile* was detected in 19% of controls (4/21) but in none of the patients treated with probiotics (P<0.05).
In conclusion, enteral administration of the probiotic bacterium *L. plantarum* 299v to critically ill patients treated with antibiotics reduced colonisation with *C. difficile*.

[Klarin et al. (2008). *Lactobacillus plantarum* 299v reduces colonisation of *Clostridium difficile* in critically ill patients treated with antibiotics. *Acta Anaesthesiologica Scandinavica* 52:1096–1102; doi: 10.1111/j.1399-6576.2008.01748.x]

ACUTE PANCREATITIS

Acute inflammation in pancreas (pancreatitis) is a severe condition, where pancreatic enzymes are activated and attack tissue which leads to ischemia and reperfusion, and bursts of oxygen free radicals which mediate inflammation and tissue injury.

Mostly pancreatitis is trigged by gallstones or alcohol. 15—20% of the patients with acute pancreatitis develop severe acute pancreatitis, with a mortality rate of 9—27%.

Pancreatitis results in a systemic inflammation that weakens the barrier function of the intestinal mucosa, and bacteria and bacterial components in the intestinal lumen can translocate into the body. In turn this leads to further aggravation of the inflammation, which hamper the barrier function of the intestinal mucosa even more. This is in risk of becoming a vicious cycle of negative events, that can result in multi-organ failure and/or sepsis, and a fatal outcome.

Mainly based on promising results in animal models, and small pilot studies on humans with *L. plantarum* as the probiotics, it has been suggested that enteral administration (tube feeding) of probiotics can help patients with pancreatitis by decreasing the bacterial translocation from the gut and suppressed systemic inflammation.

Study example: Failure to supress systemic inflammation in acute pancreatitis

In a randomised, double-blind, placebo-controlled trial 152 patients with severe acute pancreatitis enterally received probiotics (10^{10} CFU/d), twice a day for more than 28 d; 144 patients received placebo.
The brand name of the received probiotics was Ecologic 641 (Winsclowe Bio Industries, Amsterdam, Netherlands) and it consisted of a mixture of different strains of the species *Lactobacillus acidophilus*, *Lacticaseibacillus casei* (presumably *L. paracasei*), *Ligilactobacillus salivarius*, *Lactococcus lactis*, *Bifidobacterium bifidum* and *Bifidobacterium lactis* (= *B. animalis*).
The outcome of the study was negative, and nine patients in the probiotic group developed bowel ischemia, and eight died, compared with none in the placebo group.

[Besselink *et al.* (2008). Probiotic prophylaxis in predicted severe acute pancreatitis: a randomised, double-blind, placebo-controlled trial. *Lancet* 371:651—659]

The outcome in the failing study example above, where probiotics were given to patients with pancreatitis was deadly negative for some of the patients. The explanation provided by the authors of the study is that the probiotics may have increased the oxygen metabolism in the mucosa, and hence, the production of oxygen free radicals which caused a local inflammation, leading to ischemia (restriction in blood supply).

Oxygen free radicals: $O_2^{-\cdot}$ $HO_2^{\cdot}$ $^{\cdot}OH$ H_2O_2

If the authors assumption is correct, it would be of interest to know if this oxidative effect was caused by a specific strain of the mixture of different probiotic strains of different taxa that were included in the product: Of the species included into the study, *Lactococcus lactis* is the most likely one to consume oxygen due to the fact that this species has an aerobic metabolism.

However, *Lactobacillus acidophilus* is a well-known producer of hydrogen peroxide as it is lacking catalase or other means to convert hydrogen peroxide to water. This could be beneficial in some situation, for example fighting of *E. coli* on a vaginal mucosa, but can be a negative feature in a critically ill patient at risk for an over-reacting inflammatory response.

In such a situation, it is beneficial if the probiotic strain has an efficient means to convert the hydrogen peroxide to water, for example by the means of mangane-pseudocatalese as in *Lactiplantibacillus plantarum*.

Bifidobacterium bifidum is, on the other hand, unable to produce oxygen free radicals, because this is a strictly anaerobic organism unable to metabolise oxygen in any way.

The study example of pancreatitis stresses the importance in probiotic applications to use well defined bacterial strains, with well-known characteristics.

If a probiotic strain should be given to critically ill patients, the strain must have documented anti-inflammatory capacity *in vivo*. In other words, the strain must be carefully evaluated in animal inflammation-models before it is given to severely sick humans.

In this particular case, the ability of the different strains to cause oxidative stress and inflammation ought to have been more carefully evaluated beforehand.

Metabolic syndrome

LOW-GRADE INFLAMMATION

The metabolic syndrome is a combination of disorders that increase the risk of developing type-2 diabetes and cardiovascular diseases.
Factors contributing to the metabolic syndrome are:
(*i*) increased triglycerides in the blood,
(*ii*) decreased HDL cholesterol in the blood,
(*iii*) increased blood pressure,
(*iv*) increasing fasting plasma glucose,
(*v*) abdominal obesity.

The metabolic syndrome is characterised by:
(*a*) systemic, low-grade inflammation and
(*b*) increased insulin resistance.

It is unclear if the low-grade inflammation in type-2 diabetes and obesity is the cause of the condition or just a consequence of it.

Low-grade inflammation means that there is an elevation of different inflammatory markers in the blood compared to the normal levels, but without any signs of infection; the phenomenon may also be called “subclinical inflammation”.

Lipopolysaccharides (LPS) leaking out into the body from the gram-negative part of the intestinal microbiota can be the triggering factor for low-grade inflammation.

The ability of many lactobacilli to counteract *E. coli* is well known, and so is the ability of the probiotic strain *L. plantarum* 299v to mitigate bacterial translocation under diseased conditions from the gut.

It has also been demonstrated in healthy humans that *L. plantarum* WCSF1 increase the relocation of the proteins occludin and ZO-1 into the tight junction area, between duodenal epithelial cells, which is expected to result in a tighter mucosa.

The microbiota of the digestive tract is important in relation to low grade inflammation, and a favourable influence on the composition of the gut microbiota by administration of probiotics can be a strategy to mitigate the inflammation.

Ingesting probiotics can affect the composition of the resident gut microbiota, but probiotics may also have more direct effects on the immune system, and also on the permeability of the mucosa. The better the barrier function of the mucosa is the smaller is the risk of translocation of pro-inflammatory bacteria and pro-inflammatory components of bacterial origin.

Study example: Decreased cholesterol and fibrinogen in blood serum

In a small, double blinded, randomised, and placebo-controlled study, male volunteers with a slightly elevated cholesterol level in the blood consumed 400 ml ProViva daily (for 6 weeks) which corresponds to a daily dose of $2x10^{10}$ *L. plantarum* 299v. The placebo was a corresponding fruit drink without probiotics. The treatment resulted in decreased concentrations of cholesterol and fibrinogen in the blood serum of the volunteers.

[Bukowska *et al.* (1998). Significant decrease in fibrinogen and LDL-cholesterol levels upon supplementation of the diet with *Lactobacillus plantarum* (ProViva) in subjects with moderately elevated cholesterol concentrations. *Atherosclerosis* 137: 437—438]

Fibrinogen is an acute phase protein that is produced in the liver at inflammation, and is a marker for inflammation. Fibrinogen is an independent risk factor for thrombosis.

Study example: Decreased fibrinogen, F2-isoprostanes, IL-6 in blood.

In a small, double blinded, randomised, and placebo-controlled study, smokers consumed 400 ml ProViva daily (for 6 weeks) which corresponds to a daily dose of $2x10^{10}$ *L. plantarum* 299v.

The placebo was a corresponding fruit drink without probiotics. The treatment resulted in decreased concentrations of the levels of fibrinogen, F2-isoprostanes and IL-6 in the blood.
The treatment also, lowered the systolic blood pressure and decreased the serum-concentration of leptin.

[Naruszewicz *et al.* (2002). Effect of *Lactobacillus plantarum* 299v on cardiovascular disease risk factors in smokers. *Am. J. Clin. Nutr.* 76:1249—1255]

IL-6 is a pro-inflammatory cytokine that regulates the production of fibrinogen.

F2-isoprostanes are biochemical markers of lipid peroxidation and oxidative stress.

Leptin is a hormone-like messenger protein produced by fat tissue, with a complex signalling pattern, interacting with many organs and also signalling to the immune system. Leptin levels in the blood are increased at food intake, infections, fever and systemic inflammation, but are decreased at starvation, weight loss and anorexia nervosa.

In the study above, there is a correlation between leptin and fastening insulin, and also to systolic blood pressure. The finding that probiotic treatment for 6 weeks significantly lowered a marker for body fat (leptin) was a quite new and remarkable one in 2002 when the study was published.

High-energy yielding microbiota

The human intestinal microbiota contains around 10^{11} colony forming units (CFU) of bacteria per gram colonic content. Thus, the gut contains at least 10^{14} live bacterial cells, and they make up about 1 kg of biomass.

These bacteria are centrally located in the body, and interact with the host, and with food constituents in the gut. For example, they can at least theoretically influence how much energy the individual can harvest from the consumed food.

It has been suggested that the gastro-intestinal (GI) microbiota is essential for the processing of dietary polysaccharides, and that the composition of the microbiota can affect the accumulation of body fat, and of insulin resistance.

One theory is that the fat accumulation is due to an increased proportion of taxa of the phylum Firmicutes (CN: *Bacillota*), and a corresponding decreased proportion of taxa of the phylum Bacteroidetes (CN: *Bacteroidota*). In other words, a high abundance of Bacteroidetes in the gut makes the individual lean, while lots of Firmicutes makes the individual fat.

– QUESTION: Are the observed differences in microbiota really the cause of fattening?

— ANSWER: The gaining of fat can just as well be a marker of a more efficient and better functioning ecological system if the proportion of Bacteroidetes (CN: *Bacteroidota*) is low and that of Firmicutes (CN: *Bacillota*) is high.

A general weakness in this theory of the critical proportion of Bacteroidetes to Firmicutes, is that a phylum includes a wide spectrum of different taxa, each with highly different qualifications.

Firmicutes (CN: *Bacillota*) for example, includes more than 1500 different species. Some species are well-known pathogens as *Bacillus anthracis*, *Streptococcus pyogenes*, *Listeria monocytogenes* and *Clostridium difficile* (CN: *Clostridioides difficile*), causing specific and very different, but nasty diseases. And some of the bacterial taxa within the borders of Firmicutes are well-known dietary bacteria, safely ingested in high numbers with lactic acid fermented food.

Pathogens and lactobacilli have very different effects on the human body, and it can be misleading to place them under the same umbrella when discussing physiological or pathological effects.

Pro-inflammatory microbiota

Another theory to how the microbiota of the dietary tract can affect fattening is that the fat accumulation is due to an excessive amount of pro-inflammatory bacteria in the gut.

Type-2 diabetes and obesity are characterized by insulin resistance and a low-grade inflammation in the body, and the affected individuals have elevated levels of inflammatory markers in the blood; makers such as CRP (C reactive protein) and Il-6.

The concentration of leptin is also increased. Leptin can amongst several other functions be a pro-inflammatory messenger to macrophages and T-cells.

Lipopolysaccharides (LPS) originating from gram-negative bacteria in the digestive tract can act as trigger for inflammation. And a high-fat diet can increase the proportion of LPS-containing bacteria, which also can increase the risk of LPS leaking through the intestinal barrier and reaching the circulation, which in turn will lead to an increased level of inflammation.

Changes in the gut microbiota towards a microbiota with a higher proportion of LPS-producing taxa, for example from the phyla Proteobacteria (CN: *Pseudomonadota*) and Bacteroidetes (CN: *Bacteroidota*), can be caused by a high-fat diet. This can, and perhaps in combination with the high-fat diet, causes endotoxemia (LPS in the blood) which will induce systemic inflammation.

The inflammation will then initiate a process leading to fat accumulation that, overtime leads to overweight, obesity, and maybe type-2 diabetes (Figure 12).

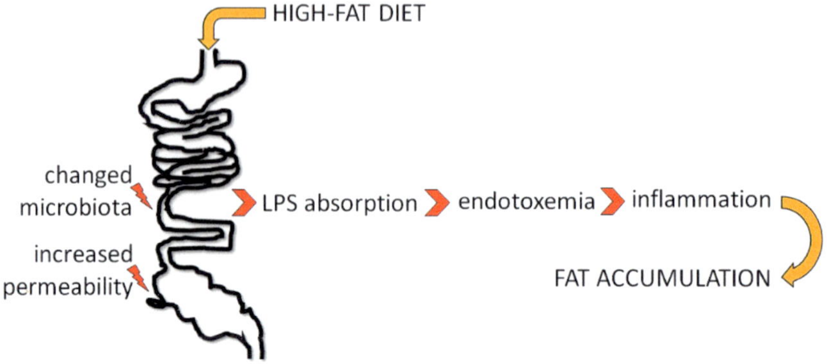

Figure 12. A leaky gut mucosa leads to translocation of pathological agents, for example lipopolysaccharides (LPS), and the consequence is inflammation. Long-term inflammation increases the risk for many diseases and may trigger fat accumulation which can lead to overweight and obesity.

The composition of the gut microbiota can have effects on the metabolic syndrome and associated disease. Fat individuals often have a less diverse gut microbiota than lean individuals, and fat ones typically harbour more *Enterobacteriaceae* than lean ones.

Use of probiotics

Disorders in the lipid metabolism can cause hypertension, and hypertension is often linked to hypercholesterolemia. There are probiotics on the market that claims that they affect the blood cholesterol.

Yoghurt supplemented with *Lactobacillus acidophilus* and *Bifidobacterium longum* has been shown to increase the beneficial HDL cholesterol.

And the intake of a sour-milk fermented with *Lactobacillus helveticus* together with the yeast *Saccharomyces cerevisiac*, was proved to decrease the blood pressure in elderly hypertensive subjects.

There is an obvious potential for probiotics aiming to counteract the dysfunctions of a low diversity microbiota, overgrowth of *Enterobacteriaceae*, and leakage of LPS through the mucosa of the digestive-tract.

Especially useful in this respect can probiotics be that strengthen the barrier effect of the gut-mucosa, and have abilities to mitigate inflammation.

Iron deficiency

Iron deficiency is primarily affecting women in reproductive age. Other risk groups are growing individuals, elderly with a poor nutrient intake, vegetarians, athletes and pregnant women.

Iron deficiency-induced anaemia during pregnancy is associated with increased risk of preterm birth, low birthweight, perinatal and neonatal mortality, and offspring neurodevelopmental disorders.

Iron deficiency can be mitigated by a diet rich in iron, but it is also important that the diet includes components enhancing the bioavailability of iron.

Ascorbic acid and meat are examples of dietary compounds that are beneficial for iron absorption; whereas calcium, polyphenols and phytic acid counteract iron absorption.

Oxidized iron has low solubility, and organisms must develop adequate approaches to compete for iron. The gastro-intestinal (GI) microbiota can, depending on composition, facilitate or obstruct iron absorption.

For example, *Escherichia coli* is efficiently mopping up available iron from the environment by the use of specific iron-chelating agents called siderophores. *Lactiplantibacillus plantarum* on the other hand, don't require any iron at all for its metabolism and biosynthesis.

Probiotics can increase the iron absorption from a meal.

Study example: Iron absorption in young females

Iron absorption from a light breakfast meal administered with or without capsules containing 10^{10} colony forming units of freeze-dried *L. plantarum* 299v, was studied in healthy, fertile female volunteers in a single blinded placebo-controlled study. The study included 42 individuals and absorbed iron was measure with a double isotope technique (^{59}Fe and ^{55}Fe). The iron absorption from a meal was significantly increased after ingestion of probiotics. Thus, freeze-dried *L. plantarum* 299v enhanced the absorption of iron when administered together with a meal with a high iron bioavailability.

[Hoppe *et al.* (2017). Freeze-dried *Lactobacillus plantarum* 299v increases iron absorption in young females: Double isotope sequential single-blind studies in menstruating women. PLoS ONE 12: e0189141. https://doi.org/10.1371/journal.pone.0189141]

Osteoporosis

Low bone mass and deterioration of micro-architectural bone tissue, with increased risk of bone fracture can be diagnosed as osteoporosis. The illness is classified as primary osteoporosis when it is a part of the natural aging process characterized by oestrogen-deficiency. The prevalence of osteoporosis in postmenopausal women is especially high.

Secondary osteoporosis occurs as part of disease pathology of other origins than ageing and oestrogen-deficiency, for example by conditions linked to an increased level of inflammation.

In osteoporosis the bone loss is due to an imbalance in the process of bone reconstruction, which can be due to many different factors.

By affecting the intestinal ecology probiotics can:

(*i*) decrease inflammation and

(*ii*) increase the absorption of calcium from the lumen.

Different probiotic strains of different species have been claimed to have effects in mitigating osteoporosis, for example, species of *Lactiplantibacillus plantarum*, *Lacticaseibacillus paracasei*, *Lacticaseibacillus rhamnosus* and *Limosilactobacillus reuteri*.

Study example: Bone loss in postmenopausal women

In a randomised, double-blind, placebo-controlled trial, early postmenopausal women were randomly assigned to receive probiotics by eating a capsule containing 10^{10} colony-forming units with a mixture of *Lacticaseibacillus paracasei* DSM 13434, *Lactiplantibacillus plantarum* DSM 15312, and *L. plantarum* DSM 15313; or a placebo, once daily for 12 months. The primary outcome was the percentage change from baseline in lumbar spine bone mineral density (LS-BMD) at 12 months. 249 participants were randomly assigned to receive the probiotic product or the placebo.
Probiotic treatment reduced the LS-BMD loss compared with placebo. The LS-BMD loss was significant in the placebo group, whereas no bone loss was observed in the probiotic-treated group.
In conclusion, probiotic treatment using a mix of the three test-strains protects against lumbar spine bone loss in healthy postmenopausal women.

[Jansson *et al.* (2019). Probiotic treatment using a mix of three *Lactobacillus* strains for lumbar spine bone loss in postmenopausal women: a randomised, double-blind, placebo-controlled, multicentre trial. *The Lancet Rheumatology* 1:E154-E162; https://doi.org/10.1016/S2665-9913(19)30068-2]

Summing up

Probiotics can improve the status of the mucosa and strengthen the mucosal barrier effect by different means. Probiotics can:
(*i*) modulate the response of the immune system;
(*ii*) strengthen the tight-junctions of the epithelial cell in the mucous membrane;
(*iii*) stimulate the epithelial cells of the mucous membrane to produce more mucin;
(*iv*) decrease the proportion of adverse bacteria in the gut;
(*v*) improve the digestion.

Different probiotic strains have different capacity to fulfil different missions.

Probiotics can improve the mucosal barrier, and probiotics can do so also at severe intestinal inflammation, or at liver injury.

In liver injury, it is especially important to block translocation of gut bacteria as those, through the portal blood will reach the liver and enhance the inflammation in the liver.

Probiotics can mitigate antibiotic associated diarrhoea, and colitis due to *Clostridium difficile* (CN: *Clostridioides difficile*).

Probiotics can mitigate rotavirus enterocolitis.

It should be noted that probiotics can act both against pathogenic bacteria and pathogenic virus; and probiotics can be active both in the small intestine (rotavirus infects in the small intestine) and in the large intestine (*C. difficille* infects the large intestine).

Probiotics are able to counteract common cold in the upper respiratory tract, that is, virus infecting the mucosa in the mouth, nose and throat.

The broad working ability of probiotics is pointing at the target for the probiotic effect is an improvement of the **status of the mucosa**, and/or an upgraded efficiency of the **immune system**. Both function of protection can be expected to have relevance for the resilience against mucosal infections, irrespectively of location and sort of infectious agent.

Functional bowel disorders are frequently occurring problems in urbanised populations. A disorder as the irritable bowel syndrome (IBS) is not linked to a specified disease. There is convincing evidence that different sort of probiotics can mitigate the symptoms of IBS, particularly pain and bloating.

When it comes to inflammatory bowel diseases (IBD), probiotics might have a role in mitigate symptoms, but are not likely to affect the actual disease. The inflammation in IBD is partly enhanced by unfavourable changes in the gut microbiota towards a higher proportion of more aggressively pro-inflammatory bacteria. Probiotics can counteract these pro-inflammatory bacteria and can presumably have a somewhat inflammation suppressing effect on the immune system.

There is evidence for the capability of probiotics to decrease markers for systemic low-grade inflammation. This is presumably, in some way or another, dependent on decreased leakage of bacteria or bacterial components through the intestinal mucosa.

It has been pointed out, that the starting point of overweight and obesity can be an increased leakage of pro-inflammatory bacteria and their components through the GI-mucosa, leading to low-grade systemic inflammation. This systemic inflammation can in turn lead to the metabolic syndrome, obesity, type-2 diabetes and cardiovascular diseases.

A high proportion of Proteobacteria (CN: *Pseudomonadota*), for example *E. coli* and other taxa of the family *Enterobacteriaceae*, in the microbiota is suggested to be the driving force of such a development.

Proteobacteria (are generally pro-inflammatory due to the association of immune-aggravating lipopolysaccharides (LPS) in the cell wall.

Concerning overweight and obesity, it has been pointed out that different compositions of the gut microbiota are differently efficient to harvest energy from indigestible polysaccharides. A high ratio of Firmicutes (CN: *Bacillota*) to Bacteroidetes (*Bacteroidota*) in the gut microbiota has been suggested to be correlated with overweight and obesity.

However, it seems unlikely that this is the cause of the fat-accumulation. A high proportion of Firmicutes (CN: *Bacillota*) in the microbiota is probably just a marker for a good intestinal ecology, and hence, an efficient energy-generating system.

It should be stressed that beneficial health effects of probiotics can be enhanced if the probiotics are combined with prebiotics. Especially valuable effects can be reached if the prebiotics exercise protective or stimulating effects on the added probiotics.

In the environment of fermented foods, the growing organism is both stimulated and protected by components in the raw food material; and several fermented foods could very well exercise prebiotic effects, but the knowledge is so far limited, and needs further clarification.

Questions:

— Do you believe in probiotics? — If yes: why?
— Do you not believe in probiotics? — If you do not believe in it: why?

— Give one example of a really convincing probiotic trial.
— Why is it good?

— Give one example of a really lousy probiotic trial.
— Why is it so lousy?

New and improved probiotics

How to administrate life

Probiotics are small living creatures, and to be fully effective they must reach the mouth and the gastro-intestinal (GI) tract of the consumer as lively as possible. This can be achieved either by:
(*i*) incorporating the probiotics in a suitable food where the organisms, at least to some degree can go on metabolizing and multiplying during storage, or
(*ii*) by different means preserving the probiotics in dense populations, in a dry state, and in a protective environment; and then present the probiotics in form of tablets, capsules or powder in sachets as a food supplement.

From the perspective of the probiotic organism, and in view of the potential of providing health beneficial effects, the incorporation of probiotics in food is superior to dried supplements by two major reasons:
(*i*) the food alternative can provide the consumer with more lively probiotics, and also
(*ii*) probiotics in considerably higher doses (for a lower production cost).

However, it should be pointed out that the supplement-strategy has commercial advantages over the food-incorporation alternative for both the producer and the retail trade, by being more easily controlled during storage, and more easily marketed.

On the other hand, a major drawback with a supplement, is the requirement of preserving techniques, regularly freeze-drying, which steeply increases the production cost.

An obstacle with incorporation of probiotics directly into food by letting the probiotic organism multiply in the food, is that it requires a food system that provide good living condition for the probiotic organism.

And at the same time, avoid ruining of the eating quality of the product during fermentation when the probiotics are multiplying and metabolizing at a high rate, but also, and even more important, during storage when the probiotics should keep their vitality without negatively affecting the eating qualities of the product. In practice, this can be a tricky balance act.

An example of how the problem can be solved is the Swedish oat-based fruit-drink, ProViva. The probiotic strain *L. plantarum* 299v is used as starter culture to ferment an oatmeal gruel under optimal growth temperature until the organism reach high numbers.

The fermented oatmeal gruel is then mixed with different fruit combination (5% [v/v] oatmeal), and packed and stored in refrigeration.

The low pH in the product that has been reached during fermentation slow down further metabolic activities of the bacteria, and this retardation in metabolism is enhanced further by the low pH in the added fruit and by polyphenols in the fruit; different fruits have different anti-microbial effects.

Fermented milk in many forms has a long tradition as carrier for probiotics, but all food constituents rich in fermentable carbohydrates can be used for probiotic production, and as probiotic food carriers; not only cereals or cereals-like constituents can be used, but also, for example, beans of legumes.

Probiotics can be based on other bacterial taxa than lactic acid bacteria. One example on a fermented product that contains living bacteria that also is claimed to have probiotic effects, is the Japanese product *natto*.

Natto has an ancient tradition, but is a product that has been converted to the buying and eating habits of modern consumers.

Natto is based on small soybeans, fermented by a strain of *Bacillus subtilis*. Before fermentation the beans are washed and soaked in water for 12—20 hours. The soybeans are then steamed for 6 hours, and after cooling they are inoculated with *Bacillus subtilis* and fermented at 40°C for 24 hours.

Afterward, the natto is aged at refrigerator temperature for up to one week to allow the development of stringiness, that is, a really sticky slime.

Industrially produced *natto* is marketed in a small 40—50 g polystyrene container, that typically complements a small bowl of rice, and are frequently eaten at breakfast in Japan.

How to control life

Life is never static, and probiotics are living organisms, that due to environmental circumstances can change characteristics with time, and perhaps even evolve.

This is a major problem from an industrial point of view, because when a probiotic organism has been isolated, described and then by scientific studies proved to have certain carefully described health beneficial effects, it becomes commercially infuriating if the organism is changing characteristics over time, and perhaps even loses its most important health carrying characteristics.

The dynamics of life is a major commercial and legislation-related problem when it comes to probiotics, and an obstacle that immediately becomes obvious for the probiotic "strain-hunter". Because, already after the first isolation from an agar-plate, many isolates tend to behave differently next time they are re-cultured on the same substrate in the laboratory.

In the search for new probiotic strain a prime condition for selecting a certain isolate for further characterization, is that it has a stable behaviour through generations of culturing.

Another condition is that the chosen isolate is easy to handle in the laboratory, and by extension in the industrial settings of production.

Otherwise, the probiotics will be too expensive to produce, and therefore, in reality unacceptable for the market of food and food-supplements.

It should always be held in mind that the chances for health beneficial effects, due to probiotic administration, increase with the number of living microorganisms in the administered dose.

The probiotic "strain-hunter" must carefully identify all promising strains down to the level of species. And when the chosen isolate scientifically has proved its health beneficial value:
(*i*) the entire genome should be sequenced,
(*ii*) the strain should be deposited in at least one international culture collection, and if possible,
(*iii*) the strain should be patented.

Furthermore, the owner of the commercial rights of the strain must find genomic measures to define the strain below the level of species; and also design a set of tests that describes the characteristics of the strain in such a way that it phenotypically describes the key characteristics of the strain that should be conserved through time and storage.

Tests to control that the strain remains unchanged over time must be listed and carefully described. These tests can be directed towards the genome, but must also reflect phenotypic characteristics that are regarded as important for the behaviour of the strain, and that is both technical performance and characteristics important for the supposed, beneficial health effects.

How to find beneficial bacteria

THE IDEA

Some absolute conditions for a successful search for a new probiotic strain is that the "probiotic hunter" has an idea of what:
(*i*) the probiotic strain should offer for health beneficial advantages;
(*ii*) type of organisms that can deliver the required beneficial effects, that is, what sorts of microorganisms (what taxa) should the search be focused on, and in what
(*iii*) environment can the required organisms be found?

THE SEARCH

The starting point for the search of new probiotic strains must begin with the picking of vivid isolates from the searching-site, for example, from the saliva of healthy individuals. This first stage of the search must be directed towards the living, intact organism and it can't be a search for genes.

Should it be of interest to search for specific genes, this search must be directed towards the captured isolates.

It must be kept in mind that the biology of living organisms is very complex indeed, and in major parts totally unknown. This implies that an initial screening after biological details, disregarding the organism as whole, is convicted to failure.

The search must start from above, from the overall picture of the biological system, and not from below, from molecular details. Thus, it should not be started with, for example, mechanistic details of the biological system, which by the way, mostly is abstractions of the biological reality.

If we make a parable with a jigsaw puzzle, and see the picture of the puzzle as the biological action of the probiotics. It is impossible to get the overall picture by reconstructing it from one single piece of puzzle.

The screening must instead start in the overall reality of the biological complexity, that is, in substantial parts of the jigsaw puzzle where the observer can get an idea of the picture.

The first stage of the biological screening is the isolation of living organisms. In other words, the starting point is living cells that are feasible to culture, and to re-culture in the laboratory.

The next stage of screening will benefit from inclusion of as large biological complexity as possible, beside of the features that are regarded as critical for the probiotic function in the human individual.

This means that the best screening system, if possible, is to use human beings as test-environment.

For example, if we want a probiotic strain able to adhere to the human intestinal mucosa as we believe that the mucosa is the site for the probiotic to act on, then we can give the research person different probiotic strains to eat and investigate which ones are most prone to adhere to the intestinal mucosa, that is, if you get an ethical permission, and can recruit enough volunteers for the screening.

Perhaps, the adhesion is not enough, if we also want to see what the probiotic candidate actually are doing on the mucosa; a study that requires surgery. Then, human research persons are then out of the question.

Instead, an animal model can be designed for the screening. Such a model is never the same as testing in humans, but anyway, it is much closer in biological similarity than artificial test-system ever can be, cell-culture tests included.

Study example: Screening in mouse after strains able to improve microbiota diversity and to mitigate *E. coli*.

The aim of the study was to implement a mouse model for *in vivo* screening of lactobacilli-strains for their ability to increase gut-microbiota diversity and to mitigate *E. coli* in the gut.
Mice were divided into different groups, two dietary control-groups and four groups administered different lactobacilli-strains. All animals were pre-treated with antibiotics, and *E. coli* in order to equalise the microbiota from the start. After 7 weeks of lactobacilli-administration, all animals were sacrificed: DNA was extracted from caecum tissue, and the microbiota composition was analysed.
The findings demonstrated that the animal model was suitable for *in vivo* studies of the effect of bacterial supplements on the gut-microbiota.

[Linninge *et al.* (2019). *Lactobacillus fermentum* and *Lactobacillus plantarum* increased gut microbiota diversity and functionality, and mitigated *Enterobacteriaceae*, in a mouse model. *Beneficial Microbes* 10:413—424]

How to prove beneficial health effects

It is difficult

It is difficult and requires a lot of resources to prove an effect of a drug against a specific disease, but it is even more difficult to prove health beneficial effects of dietary compounds such as probiotics, in non-diseased subjects.

It is often the prevention of a disease that shall be proved, and randomised placebo-controlled studies ought to be run for long time periods in order to prove the preventative effect. However, when it comes to probiotics it may be both economically and practically problematic to run placebo-controlled trial for sufficient long time periods.

Instead, cohort studies backed up with animal studies can be an accessible way to prove the causal connection. On the other hand, if the probiotic-strain is powerful enough to give short time effects, it is feasible to apply placebo-controlled studies for the testing of probiotics, especially if the probiotic company has sufficient economical resources at hand.

The value of the evidence from placebo-controlled studies is much dependant on the quality of the endpoints of the study. The best endpoints are those of improvement or prevention of dysfunctions.

The value of evidence in the form of improvement of biomarkers is much weaker, and the relevance of biomarkers can always be questioned. However, some biomarkers are more established and reliable than others. For example, cholesterol in the blood is an established biomarker in the metabolic syndrome for predicting the risk for cardiovascular diseases.

There are many immune-related compounds that at least theoretically can be used as biomarkers for the status of the immune system and as such giving hints about the health-status. On the other hand, there are often different opinions in the scientific community about how these immunological biomarkers should be interpreted.

Evidence based

Probiotics are living microorganisms with scientifically demonstrated health beneficial effects when administered to the body. A crucial question is then, how should these health beneficial effects be proved?

A prerequisite for the Natural sciences is that statements must be evidence-based. Evidence-based means that rules for proving a statement should be set up before-hand, and then followed.

It is not in line with the idea of Natural science to have a preconceived explanation and then search for proofs to support this explanation, and in the process neglecting data speaking against it.

It is important in evidence-based science that any trained scientist should come to the same conclusion when mimicking the study and following the rules that have been set up.

In evidence-based medicine, evidence obtained from different types of studies are ranked differently. Highest value in proving the effect of a certain treatment in medicine is:
(*i*) the ***randomized, placebo-controlled trial*** (often abridged, RCT),
(*ii*) the ***cohort study*** and then following in order,
(*iii*) ***case control study***,
(*iv*) ***case report***,
(*v*) ***expert opinion***,
(*vi*) animal models and
(*vii*) *in vitro* studies.

It may seem odd that expert opinions are ranked higher than animal studies. However, this is medicine where the object always is the patient and the expert in this case is a trained and experienced physician that during the daily work comes in contacts with a lot of patients and different clinical situations, that is, the expert opinion is based on empirically gained knowledge from observations of the reality.

In many other fields of Natural sciences, as for example nutrition, a well-designed animal study must be regarded to have a considerably higher explanation-value, than the opinion of an individual scientist or nutritionist.

In the expression ***randomized, double blind, placebo-controlled trial***; "double blind" implies that all subjects in the study get a product where they are unaware if it contains the active compound or not, that is, the administrated products are coded and not only the study-subjects but also the operators of the study, and the ones performing the analyses are unaware of which individuals that receive the active product and which ones that receive the placebo.

Placebo is the seemingly similar product as the test-product, but without the active compound.

"Randomised" implies that test persons at random are divided into different test groups, that is, it is not allowed to select certain persons for certain test groups.

Sometimes so called ***cross-over*** is applied, meaning that after a certain run-time of the study, the volunteers that got the placebo instead get the active product, and vice versa.

In a so called ***open-label trial***, both the researchers and participants know which treatment that is being administered. Open-label trials may be appropriate for comparing two, very similar treatments to determine which one of them that are most effective. An open-label trial can still be randomized.

A “cohort” is a group of subjects who share a common characteristic or experience within a defined period. A ***cohort study*** (sometimes also called “panel study”) is an observational study.

Cohort studies can either be conducted prospectively (in the future) or retrospectively (looking backwards) from archived records.

A ***case-control study*** identifies factors that may contribute to a condition by comparing subjects who have that condition (the cases), with subjects who are without the studied condition, but otherwise similar (the controls).

Case report is used in medicine to describe, in detail the symptoms, signs, diagnosis, treatment, and follow-up of an individual patient.

All trials have an ***endpoint*** (end-point) or a primary outcome, that is, a mark of termination or completion. In clinical or nutritional trials, the endpoint refers to the sign that defines the final outcome of the trial.

Evidence-based medicine can be defined as the integration of high-quality research-evidence with clinical expertise and patient values. The concept of evidence-based medicine has been transferred to the field of nutrition, that is, evidence-based nutrition.

Also in nutrition, including probiotics, the randomized, placebo-controlled trial (RCT) has the greatest weight of proof, because RCT is the only design that permits strong causal (implying an effect) inference (the act of conclude from evidence).

Koch's postulates

The classical way to prove the pathogenic ability of a microorganism and link it to a particular disease is to follow the different steps in Koch's postulates that were set up by pioneer microbiologist and infection physician Robert Koch in 1884:

1) The microorganism can always be found in association with a particular disease, but not in healthy individuals.

2) The microorganism can be grown in the laboratory in pure culture.

3) This pure culture will produce the same disease when inoculated into a new, susceptible individual.

4) The microorganism can be recovered from the infected, diseased individual and grown again in pure culture in the Laboratory.

Should the postulates of Koch be converted for the proving health beneficial effects of a probiotic strain it can be:

1) The strain can be found in association with healthy individuals.

2) The strain can be grown in the Laboratory in pure culture.

3) This pure culture will exercise health beneficial effects when administrated to a new subject.

4) The strain can be recovered from the inoculated subject and grown again in pure culture in the Laboratory.

There are some general difficulties to apply the modified Koch's postulates to probiotics: One obstacle is to isolate and identify the probiotic strain from the mouth or the digestive tract where a wide range of similar, related organisms may be present. This requires a method where individual bacterial strains can be identified and distinguished from all other bacterial strains of the same species.

In contrast to true pathogenic organisms which only are found in diseased individuals, probiotic bacteria are not necessarily limited to healthy individuals. They may very well be present in diseased individuals, where the probiotic bacteria do not manage to cure the disease, but the probiotics are not the cause of the disease.

And last but not least, from a medical point of view it is relatively easy to measure how a patient is cured from a disease, or getting a disease after being infected with a pathogenic agent. It is much more difficult to prove improvement of health in a non-diseased individual, or the prevention of future diseases, by the consumption of a specific probiotic strain of microorganisms.

Medicine versus Nutrition

In general. In Medicine, the endpoint is a disease or at least a dysfunction with a diagnosis, and the treatment aim to counteract the disease or dysfunction. If the treatment involves a substance administrated orally, it means that the subject will be eating a drug.

Usually, a drug will be a defined and powerful substance that is active in low concentrations. Hence, the effects of the drug can be noted after relatively short time of treatment, and if a randomised placebo-controlled trial (RCT) is to be designed for proving the efficiency, the individuals in the placebo group will never come in contact with active compound.

This is in sharp contrast to the situation in Nutrition, where the endpoints are improvement of health-status or prevention of future dysfunctions or diseases. Furthermore, the test substance (the active food component) should preferably be tested in primarily healthy individuals.

Food components that are supposed to be wholesome without side effects are usually less powerful than drugs, and the health-improving or ill-preventing effects are often not seen at first, but need prolonged exposure periods to be made visual.

If a RCT is set up, it is in Nutrition often difficult to have a placebo group where the individuals never come in contact with the active compound as it is a dietary compound, and test persons in the placebo group can't stop eating, and must include essential nutritional components in their diet.

The problem becomes especially troublesome if the study should be run over long time-periods.

From birth to death. Nutritional factors can affect the health status from birth to death. However, when a disease has emerged, it is the responsibility of Medicine to mitigate developed dysfunctions.

The metabolic syndrome can be taken as an example Figure 13). During the life span an individual can acquire the metabolic syndrome, which will increase the risk for diseases as non-alcoholic fatty-liver disease, type 2 diabetes and cardiovascular diseases.

Nutritional factors can have long-term effects and decrease the risk of developing the metabolic syndrome, and they can mitigate the symptoms of the metabolic syndrome, which in turn will decrease the risk for developing disease. However, when a specific disease hit an individual, the disease will primarily be treated with medicines.

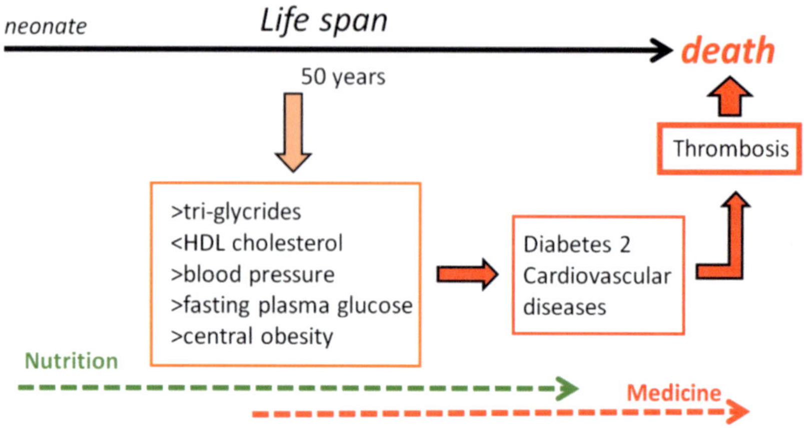

Figure 13. Nutrition and Medicine are scientific disciplines applicable at different stages of the metabolic syndrome.

Medical drugs are active at low concentrations and show effects after relatively short exposure time. In contrast, if we believe that supplement of a certain dietary compound, let's say multivitamins, through life decrease the risks for developing high blood pressure in the age of 55, how should that be evaluated by RCT?

It becomes too complicated, too expensive, and ethically totally impossible to have a RCT going for years, where the research persons in the placebo group are denied important food components as, for example, vitamins.

Instead of a RCT, it is now better to set up a cohort study. This in order to see if there are any negative correlation between the consumption of the dietary compound (in this case multivitamins) and the disease or dysfunction (in this case high blood pressure).

If such correlation is found, it remains to be proved that the daily consumption of multivitamin also is the actual cause for the blood pressure reducing effect, and not just a consequence caused by some other factor than the vitamins.

Such causal evidence can then be collected with the use of animal models. On the other hand, this proof of concept is weaker than a RCT, but in many cases is the only feasible way.

With this in mind, results from short-time RCT within the field of nutrition that in spite of all the muddling circumstances actually points at significant improvements in health status should not be easily dismissed, especially if several studies point in the same direction.

Probiotics can be regarded as dietary supplements and probiotics are primarily aiming to improve the health-status and prevent dysfunction and diseases. A general problem in evaluating the probiotic effect in RCT is that the same bacterial strain or strains with similar effects as the test-strain can be spontaneously present in some of the test subjects without any administration of the active product.

The probiotic strain or similar strains can also occur spontaneously in lactic acid fermented foods that the subject might be eating during the study time.

Test persons can also be contaminated with the test-strain from other individuals by direct contact.

Furthermore, it is impossible in a RCT aiming to test a specific probiotic strain to get rid of the disturbing influence of the test person's own microbiota of resident bacteria, and it is extremely difficult to exclude all influence of environmental bacteria contaminating the test person during the study time.

Animal models

Animal models are helpful means for comparing the causal effects between different probiotic strains *in vivo*, and for clarifying mechanisms of action, and to certifying safety.

Mice and rats are the two most frequently used animal species for *in vivo* tests.

The interpretation of test results from animals can never be directly transferred into human settings, and the interpretation must always be done with considerably caution. But still, an animal model is always much closer to the biological reality than *in vitro* tests.

QUESTIONS:

— If you should launch new probiotics of your own choice on the market, how should that product look like?

— How would your strategy be for finding a new probiotic strain?

Student's guide

to microb-biology

Organizing organisms

Bacteria

The domain, Bacteria consist of small, ancient life-forms, which are found in nature as individual cells, cells aggregated together in lumps or cells forming biofilms on surfaces. The bacterial cell (the bacterium) has a single circular DNA chromosome within the cytoplasm, but no nucleus; and the bacterial cell lacks intracellular organelles.

In short, bacteria can be described as an ancient domain, of small, usually unicellular organisms.

Bacteria are organisms that are characterized by their small and tiny size. One ordinary bacterial cell enclosed in one ml of water correspond, in chemical terms, to a concentration of one ppt (ppt = parts per trillion, that is 1 per 10^{12}).

Bacteria make up one of the three domains of life-forms (organisms); the other two are Archaea and Eucarya (or Eukaryota). The latter includes amongst others animals, plants and fungi.

The former, Archaea, are morphologically similar to Bacteria, but are often adopted to extreme environments, or have specialised metabolic pathways. For example, some archaeal taxa are given off methane as catabolic end-product.

Archaeal taxa seem to lack most sort of pathogenic ability.

Systematics

In biology, the division and organizing of different sort of organisms is handled within the discipline **systematics**, that is, systematics is the science about organism's sort and diversity, and about all sorts of relationships between them.

Synonymously, the word **taxonomy** is often used. However, to be strict, taxonomy [from the Greek táxis (order, formation), and nómos (law)] is the science of classification.

Classification is the division into groups in a hierarchical fashion, according to similarity and relationship.

Biological classification always proceeds stepwise from the lowest hierarchical level.

Classification is a fundamentally different from identification, even if the two concepts, most unfortunately, sometimes are mixed up.

Identification is the comparison of an unknown object with known ones, and proceeds in general terms in the opposite direction compared to classification, that is identification goes from the hierarchical top, down to the bottom.

Generally biological classification organizes organisms in a hierarchical fashion, starting from the individual organism and going upwards lumping them together in different hierarchical categories (Figure 14).

Domain
⇧
Phylum
⇧
Class
⇧
Order
⇧
Family
⇧
Genus
⇧
Species
⇧
Subspecies
⇧
Strain

Figure 14. The different hierarchical levels in bacteriology. Arrows indicate the direction of classification.

Bacteriological classification is currently mostly built around the supposed evolutionary pathway of organisms, that is, the **phylogeny** of the organisms.

The word "phylogeny" was originally launched by the German zoologist Willi Hennig in his book *Grunzüge einer Theorie der Phylogenetischen Systematik* (1950). He criticised the mixed phenetic (classifying according to similarity) and evolutionary view on classification. Hennig wanted to purify the evolutionary way of thinking, and he made up working-rules based strictly on evolutionary principles. He introduced the terms "phylogeny" (phylogenetic) and "cladogram" (clados is branch in Greek). The latter is a diagram (tree) of phylogenetic relationships between taxa.

The official forum for bacterial **nomenclature** and the publication of new bacterial taxa from the rank of subspecies to that of phylum is the *International Journal of Systematic and Evolutionary Microbiology*. This journal publishes research papers and establishes novel bacterial names.

New names are summarised in a notification list. Each monthly issue also contains a compilation of validated new names (the validation list), that is names and descriptions of taxa that have been previously published in other scientific journals or books are listed in the *International Journal of Systematic and Evolutionary Microbiology*, provided the descriptions and naming have been performed correctly.

Publications describing new bacterial taxa, and validation of descriptions, requires that a type strain of the taxon in question have been deposited in two recognised public strain collections in different countries.

The ICSP's International Code of Nomenclature of Prokaryotes states that all new bacterial names must be published or indexed in the *International Journal of Systematic and Evolutionary Microbiology* to be deemed valid.

All validly published names in bacterial systematic, with the original references, together with the hierarchical structure for bacterial taxa from phylum to species can be seen on a website with the title: LPSN - List of Prokaryotic names with Standing in Nomenclature (https://www.bacterio.net/).

This site shows the whole bacterial classification system and includes all taxonomical hierarchies from phylum to subspecies. It also points out the correct name (CN) of the taxa, what synonyms there are, and the publication where the taxum was describes for the first time.

Phylum

Frequently occurring phyla

As the number of bacterial species gradually increases, new ones are defined, and along with the increasing number of different species also the number of different taxa at higher hierarchical ranks increases, and so do the number of the highest hierarchical rank, that is, phylum. Or as it is called in more traditional terminology, division.

Examples of phyla (plural for phylum) frequently occurring in the human digestive tract are:
Actinobacteria (correct name [CN]: *Actinomycetota*; gram-positive),
Bacteroidetes (CN: *Bacteroidota*; gram-negative),
Firmicutes (CN: *Bacillota*; gram-positive),
Fusobacteria (CN: *Fusobacteriota*; gram-negative),
Proteobacteria (CN: *Pseudomonadota*; Gram-negative), and
Verrucomicrobia (CN: *Verrucomicrobiota*; gram-negative).

Bacteroidetes

The phylum Bacteroidetes (CN: *Bacteroidota*) includes several classes, orders and a lot of families.

Bacteria classified as the phylum Bacteroidetes (CN: *Bacteroidota*) consists often, when found in ileum and colon, of the gram-negative genus *Bacteroides*.

The genus *Bacteroides* includes many different species but some of the more, frequently occurring ones in the digestive tract are *Bacteroides vulgatus*, *Bacteroides uniformis*, *Bacteroides thetaiotaomicron*, *Bacteroides ovatus*, *Bacteroides stercoris*, *Bacteroides caccae*, *Bacteroides putredinis*, *Bacteroides merdae*, *Bacteroides fragilis*, *Bacteroides intestinalis*, *Bacteroides pectinophilus*, *Bacteroides finegoldii*,

Bacteroides eggerthii, *Bacteroides capillosus*, *Bacteroides dorei* and *Bacteroides xylanisolvens*.

B. fragilis is the type species of the genus *Bacteroides*.

Closely related species to *B. fagilis* are *B. caccae*, *B. distasonis*, *B. eggerthii*, *B. merdae*, *B. ovatus*, *B. stercoris*, *B. thetaiotaomicron*, *B. uniformis* and *B.vulgatus*.

B. fragilis is also the most aggressive *Bacteroides* species, and the one that most frequently is involved in secondary infections of the abdominal.

Some *B. fragilis* strains can produce toxins. Toxinogenic strains have been suggested to be involved in the initiation of colorectal cancer.

It has been shown that patients suffering from inflammatory bowel disease (IBD) have a higher proportion of *B. fragilis* on the intestinal mucosa than healthy individuals.

Bacteroides spp. often produce acetate and butyrate in the digestive tract, and sometimes also propionate.

Firmicutes

The gram-positive phylum Firmicutes (CN: *Bacillota*; Gram-positive bacteria with low guanine + cytosine ratio of the DNA [G+C <55%]) consists of even more classes and orders than Bacteroidetes (CN: *Bacteroidota*), and a lot of different families. Firmicutes (*Bacillota*) includes well-known genera as, for example, *Clostridium*, *Eubacterium*, *Ruminococcus*, *Faecalibacterium*, *Peptostreptococcus* and *Dialister*.

Firmicutes (*Bacillota*) occupies a dominating position in the digestive tract, typically represented by genera as *Clostridium*, *Ruminococcus* and *Eubacterium*.

The genus *Clostridium* has historically been somewhat of a taxonomic problem: It was in the past based on phenotypic traits, and was defined as strictly anaerobic rods that were able to form endospores (endospore = a single spore formed in the interior of the bacterial cell).

This definition was shown to be way to inexact for definition of a genus, and when classification of bacteria based on the 16S rRNA gene was introduced, *Clostridium* had to be redefined based on phylogenetic criteria.

Actinobacteria

The phylum Actinobacteria (CN: *Actinomycetota*; gram-positive bacteria with a high guanine + cytosine content [G+C >50%]). The gram-positive genus *Bifidobacterium* belongs to Actinobacteria (CN: *Actinomycetota*).

Proteobacteria

The phylum Proteobacteria (CN: *Pseudomonadota*), is a major phylum of gram-negative bacteria. The phylum includes a wide variety of pathogenic genera, but also free-living, non-parasitic, as for example, bacteria responsible for nitrogen fixation.

Their outer membrane is mainly composed of lipopolysaccharides (LPS).

Many are motile by flagella, but some are nonmotile, or gliding.

Proteobacteria (CN: *Pseudomonadota*) have a wide variety of different metabolisms. Most are facultatively or obligately anaerobic, chemolithoautotrophic, and heterotrophic, but numerous exceptions occur.

A variety of genera, which are not closely related to each other, convert energy from light through conventional photosynthesis or anoxygenic photosynthesis.

Proteobacteria (CN: *Pseudomonadota*) are divided in different classes, for example: *Alphaproteobacteria*, *Betaproteobacteria*, *Gammaproteobacteria* and *Deltaproteobacteria*.

Food related taxa

Most gram-negative, food associated bacteria as, for example, the genera *Pseudomonas*, *Psychrobacter*, *Aeromonas* and the family *Enterobacteriaceae* belong to the phylum **Proteobacteria** (CN: *Pseudomonadota*), and the class: *Gammaproteobacteria*.

Most gram-positive bacteria associated to food belongs to the phylum Firmicutes (CN: *Bacillota*), for example the genera *Lactobacillus*, *Bacillus*, *Staphylococcus*, *Enterococcus* and *Lactococcus*.

Direct gene identification

In order estimate the proportion of dominating bacterial taxa (groups) in environmental samples, direct gene identification can be used. All bacterial genes in a sample are extracted and then identified (or classified).

Direct gene identification has up to relatively recently been directed towards the 16S ribosomal RNA (rRNA) gene, currently, so called next generation sequencing or massively parallel sequencing gradually are taken over in order to obtain more accurate and complete picture of the microbial community composition.

The term **metagenomics** is frequently used as a label for studies where more or less all the genetic material is recovered and identified directly from environmental samples.

Species

The species is the most important category (taxum) in biological taxonomy. The species defines a group of organisms that between themselves has a high genomic similarity (the genome is the genetic material of the cell).

The species is in biological systematics named by the genus name followed by the species epithet, for example *Lactiplantibacillus plantarum* where *Lactiplantibacillus* is the genus and *plantarum* is the species epithet.

The genus is the hierarchical level above that of species, that is, related species are gathered in a genus (genera in plural). For each genus, there is a **type species** which defines the genus in question.

The aberrations "sp." and "spp." after a genus name stands for one unnamed species (species in singular) and several unnamed species (species in plural), respectively. For example, *Lactobacillus* sp. (sp. = species [one single species]), and *Lactobacillus* spp. (spp. = species [several species]).

The total number of known bacterial species is around eleven thousand (steadily increasing in numbers), that is, the number of bacterial species that have been formally described, and given a species name of their own, that is a species name that has been validly published in the scientific literature.

The species concept differs between different biological disciplines, and the genomic span between different populations (strains) of bacteria within the same species is much wider than between individuals in a botanical or a zoological species.

This is due to the fact that the morphology of bacteria is much less varied (and more unstable) then for plants and animals, and at the same time, has morphology, historically, been the starting-point in all biological classification. Morphology is especially dysfunctional in the classification of bacteria.

The current bacterial definition of a species is based on genomic methods, but so far these methods had led to the construction of relatively heterogeneous species. The generally accepted standard-method for species definition in bacteriology has for long been **DNA:DNA-hybridization**, that is to compare the similarity between the total genomic DNA between organisms in a par-wise fashion.

Unfortunately, this method is both work-intensive and somewhat unreliable from the technical point of view. The DNA:DNA hybridization homology of different bacterial isolates has to be performed par-wise, which is inconvenient and work intensive.

It is generally accepted that isolates with a relative ratio of binding of ≥70% DNA:DNA-homology, at optimal and stringent re-association temperatures, are regarded to belong to the same species. Optimal temperature is a temperature 25°C below the melting point of the DNA. Stringent temperature is the temperature 15°C below the melting point of the DNA.

The **16S rRNA gene** sequence similarity of 99% is often regarded as a species definition. But, even if the similarity in the 16S rRNA gene is 100% between strains, it is not absolutely certain that the isolates belong to the same species. For example, the 16S rRNA gene sequence of the species *Lactiplantibacillus plantarum* and *Lactiplantibacillus pentosus* are 100% similar.

A higher resolution than comparing 16S rRNA genes can be achieved when the whole genome (or at least a major part of the genome) is taken into account.

Average nucleotide identity (ANI), is one whole genome similarity matrix that is used in order to define a species. ANI represents the average nucleotide identity of all orthologous genes shared between any two genomes.

Orthologous genes, are genes in different species that originate by vertical descent from a single gene of the last common ancestor, that is horizontally transferred genes are omitted from the ANI-comparison.

Strains (genomes) with ANI similarity >95% is regarded to belong to the same species.

Strain

The strain is the lowest hierarchical category (or level) in bacterial taxonomy. Theoretically, bacterial cells with identical DNA-sequences of the genome form a strain.

In practice, the strain definition is dependent on what method that is used for the genomic comparison. In the end, it will be the ability to demonstrate genomic differences between isolates that determines the strain. The better the resolution capacity of the method is, the more precise will the strain definition become.

The ultimate method for strain-identification is the full DNA-sequence of the genomic DNA, that is the sequence of the whole genome of the cell (plasmids and other mobile units included). Full genome-sequencing is still a relatively expensive method, even if the cost steadily decreases due to technical improvements.

A more pragmatic method for strain-definition is to use "Restriction Endonuclease Analysis" (REA) of the total genome, with the use of frequently cutting endonucleases.

But, more commonly used is "Pulsed Field Gel Electrophoresis" (PFGE), where only relatively large pieces of DNA are separated. Thus, the applied restriction endonuclease in PFGE must be an enzyme that cut the chromosomal DNA in relatively large fragments. This means that PFGE compared to REA has less capacity to visualise small differences in the genome; in REA more frequently cutting endonucleases can be applied, and the resulting band-pattern will then reveal more information. But, from a technical point of view, PFGE is technically less complicated to run than REA.

If a strain should be used commercially, for example as probiotics, it is recommended that it is defined by sequencing of the whole genome. However, there are situations in control and in research where large numbers of isolates must be screened for strain-identity, and here is full genome-sequencing far too expensive and time consuming.

Then PFGE is preferably used, or even more simplified methods, for example "Randomly Amplified Polymorphic DNA" (RAPD). This PCR-based method is easily performed, and can relatively quickly be run for large numbers of isolates. A drawback with RAPD is the low reproducibility between different PCR-machines and laboratories. Depending on the actual species and laboratory protocol, a RAPD-type can be more or less close to the strain-level.

Type strain is the strain that represents the living key of the species. When a new species is scientifically described for the first time and validly published, the scientist must select one isolate that is typical for the described new species, and deposit this strain in two internationally recognized culture collections.

The type strain for the species *Lactiplantibacillus plantarum* is written *L. plantarum* ATCC 14917^{T}, where ATCC stands for "American Type Culture Collection", 14917 is the Culture collection's identity-number, and T marks out that this strain is the type strain for the species in question.

The type strain can also be purchased from other international culture collections, but then will both the letter code and the digit number be different; the T for type strain will remain.

An example of the full, systematic categorizing of three different strains is given in Table 2.

Table 2. Taxonomic position of three different strains belonging to three different phyla. T after the strain number marks that the strain is the type strain of the particular species.

Phylum	Proteobacteria (*Pseudomonadota*)	Firmicutes (*Bacillota*)	Actinobacteria (*Actinomycetota*)
Class	*Gammaproteobacteria*	*Bacilli*	Actinobacteria

Order	*Enterobacterales*	*Lactobacillales*	*Bifidobacteriales*
Family	*Enterobacteriaceae*	*Lactobacillaceae*	*Bifidobacteriaceae*
Genus	*Escherichia*	*Lactiplantibacillus*	*Bifidobacterium*
Species	*Escherichia coli*	*L. plantarum*	*B. animalis*
Strain	*E. coli* ATCC 11775^T	*L. plantarum* 299v	*B. animalis* BB12

Isolate

The word, **isolate** is often used in descriptions of practice in microb-biology. An isolate, in the bacteriological sense, is a population of bacterial cells that are originating from one single cell, which has been isolated and multiplied with the help of laboratory-technics. This type of population is called a **pure culture**.

When characterizing bacteria, it is never individual cells that are characterized. It is populations of living cells in a pure culture, or genes amplified to multiple copies, for example, by "polymerase chain reaction" (PCR) that are studied.

When a pure culture is studied for taxonomic purpose, it can be characterised with both phenotypic and genotypic methods (Figure 15).

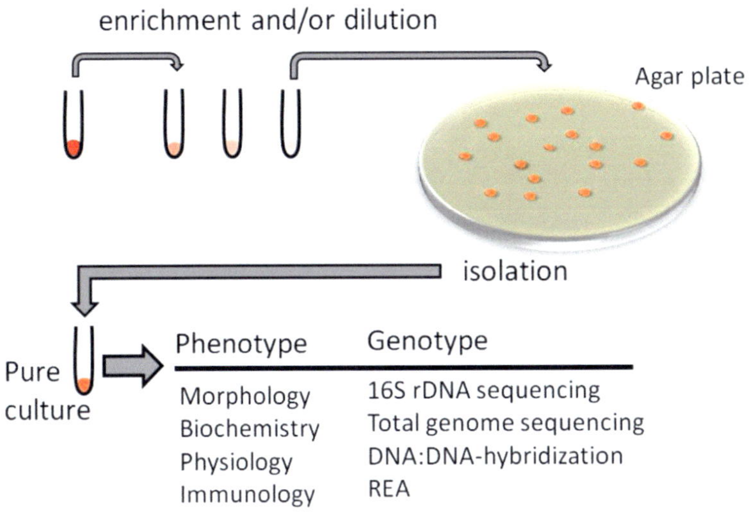

Figure 15. Viable count for bacteriological characterization, classification or identification.

When a dilution series followed by plating on agar is used for enumeration of live microorganisms in a sample, it is called **viable plate count**, which results in a number of "colony forming units" (CFU).

The phenotypic characteristics (Figure 15) are equal to the expressed characteristics of an organism, or in other words, the characteristics that are visualised in reality. This is in contrast to the genomic characteristics, that is the genotype, which are the structure, or the construction of the genome.

Examples of different types of phenotypic characteristics are morphological, physiological, biochemical, structural and immunological ones. Phenotypic characteristics are crucial as they reflect the reality but they can from a taxonomic stand point easily be misinterpreted. Morphology is especially unreliable for bacteriological classification, but can be usable for identification. In fact, traditional phenotypic features have been the most used way for of bacteriological classification and identification in the past.

Examples of phenotypic key tests often used for identification are the gram-reaction, morphology (for example, cell form in the microscope as rods or cocci, and colony appearance on an agar plate), fermentation ability of different carbohydrates, the oxidase reaction ("oxidase" is in this respect actually aiming at cytochrome c and cytochrome oxidase that is the same as cytochrome a), the catalase reaction, mobility, and growth under different specialised conditions (for example, on Rogosa agar plates).

Gram-staining was developed by Christian Gram (1884) and has proven to be an important phenotypic character that is reflected in the phylogenetic relationships between bacteria. Most members of the phylum Firmicutes (CN: *Bacillota*) are gram-positive while all other phyla in general are gram-negative.

The cell wall of gram-positive bacteria has one layer of cell membranes, and a thick layer of peptido-glycans.

In contrast, the cell wall of gram-negative bacteria has two layers of cell membranes and, in between, a thin layer of peptido-glycans.

Gram-staining is technically tricky to perform, and an alternative way is to treat a colony with concentrated KOH and record if the cell-walls break. Gram-negative bacteria are more subjected to lysis.

[Gregersen (1978). Rapid method for distinction of gram-negative from gram-positive bacteria. *European Journal of Applied Microbiology and Biotechnology,* 5: 123-127]

Many gram-negative bacteria have lipopolysaccharides (LPS) in the outer membranes of the cell wall. LPS is regarded as an endotoxin and will activate the human immune system if it comes in contact with it. Many types of bacterial LPS are strongly proinflammatory, for example, LPS from *Escherichia coli.*

If LPS comes into the circulation (blood stream) it causes immunological reactions that depending on concentration can result in fever (in concentrations of nanograms of LPS), fall in blood pressure, internal bleeding, coagulation of the blood and irreversible chock (death).

Taxa frequently used as probiotics

Genera derived from the genus Lactobacillus

Taxonomic history

The genus *Lactobacillus* belongs to the family *Lactobacillaceae* which is included in the order *Lactobacillales*, that in turn is included in the class *Bacilli* (or Firmibacteria), under the phylum Firmicutes (CN: *Bacillota*).

A traditional, short description of the cells of the genus *Lactobacillus* is that they are (*i*) rod-formed; have a (*ii*) gram-positive cell-wall; are (*iii*) unable to form endo-spores, and have (*iv*) the ability to ferment carbohydrates into lactic acid.

Lactobacillus is spontaneously dominating most traditional lactic acid fermented foods. But *Lactobacillus* can also be associated to mucous membranes of the oral cavity, digestive tract and vagina of both animals and human beings.

The genus *Lactobacillus* as it stood until 2020* included 261 names of different species and subspecies. More than 80 of these species have been certified for safe, technological, and beneficial use by the European Food and Feed Cultures Association.

In 2017, 36 species had qualified presumption of safety (QPS) status according to the European Food Safety Authorities (EFSA), and 12 species had been generally recognized as safe (GRAS) according to the United States Food and Drug

Administration (FDA). Generally speaking, all *Lactobacillus* spp. spontaneously occurring in traditional, lactic acid fermented foods in high numbers can be regarded as safe for consumption.

*The genus *Lactobacillus* was 2020 divided, and twenty-three new genera were constructed out of the former *Lactobacillus*.

Food microbiologists have since long divided the different lactobacilli into two functional groups based on what end-products that they form during fermentation of hexoses, or to be exact: fermentation of glucose, at oxygen limitation.

If lactic acid is the only end-product from glucose, the lactobacilli are said to be homofermentative, but if also carbon dioxide and ethanol and/or acetic acid is formed together with the lactic acid, the lactobacilli are said to be heterofermentative. Thus, homofermentative lactobacilli converts anaerobically glucose to lactic acid only, while heterofermentative lactobacilli converts glucose to lactic acid, carbon dioxide, ethanol, and depending on prevailing circumstances, to acetic acid.

Fermentation, is originally defined as catabolism/growth in absence of oxygen.

The functional division in homofermentative and heterofermentative lactobacilli becomes a bit more complicated by taking the fermentation of pentoses into account: Some homofermentative lactobacilli have the ability to ferment pentoses, and they ferment pentoses to lactic acid and acetic acid. Some of these can also ferment malic acid to lactic acid and carbon dioxide, and/or, ferment citrate to diacetyl, acetoin and carbon dioxide. These basically homofermentative lactobacilli are said to be facultatively heterofermentative while the true heterofermentative lactobacilli are said to be obligatory heterofermentative.

Facultatively heterofermentative lactobacilli growing on glucose in the presence of oxygen (aerobic conditions) produce beside lactic acid also acetic acid, diacetyl and acetoin.

Lactobacilli are divided into three functional groups:

1) Obligatory homofermentatives

2) Facultatively heterofermentatives

3) Obligatory heterofermentatives

The functional division of lactobacilli into
(*i*) obligatory homofermentatives,
(*ii*) facultatively heterofermentatives, and
(*iii*) obligatory heterofermentatives is of pragmatic relevance for food technologists as the catabolic end-products affects the food quality. Carbon dioxide can be a food technological problem, especially in connection with packed products, where too much acetic acid in the product gives the product a sticky taste. Ethanol has its specific qualities, unwanted in some products but most wanted in others.

The type species of the genus *Lactobacillus* is the species *Lactobacillus delbrueckii*. Each bacterial genus is defined by a type species, and the type species is the starting point for the description, not only of the species but also of the genus in question. *L. delbrueckii* can be characterized as obligatory homofermentative.

The homo-/heterofermentative concept has poor taxonomic relevance.

The former genus *Lactobacillus* (valid before 2020) was vast, and the genomic variation between different *Lactobacillus* species was too large for a genus. Therefore, it was well known that the former *Lactobacillus* genus ought to be split into several different genera, and only *Lactobacillus* spp. relatively closely related to *Lactobacillus delbrueckii* should remain as *Lactobacillus*. The problem was how this split should be performed.

Taxonomic efforts to understand the phylogenetic relationships within the genus *Lactobacillus* had up to 2020 mainly been directed towards comparative analysis of 16S ribosomal ribonucleic acid (rRNA) gene (rDNA) sequences, but also other genes had been targeted, and so had whole-genome alignments. All analyses showed an extreme divergence of the *Lactobacillus* genome and support the recognition of subgeneric division. Furthermore, to make it even more complicated, several *Pediocccus* spp. are closely related to *Lactobacillus*.

It was before 2020 suggested, on phylogenetical grounds that the genus *Lactobacillus* should be divided into ten subgroups, presumably on the hierarchical level of genus. *L. delbrueckii* and similar species made up one group of true lactobacilli (*Lactobacillus*), including species as *L. acidophilus*, *L. johnsonii* and *L. crispaus*.

Examples of other groups were the *L. plantarum*-group, including species such as *L. pentosus* and *L. paraplantarum*; the *L. casei*-group, including species such as *L. paracasei* and *L. rhamnosus*; the *L. reuteri*-group, including species such as *L. fermentum* and *L. vaginalis*; and the *L. buchneri*-group, including species as, *L. brevis*, *L. kefiri* and *L. kimchicus*.

[Salvetti *et al.* (2018). Comparative genomics of the genus *Lactobacillus* reveals robust Phylogroups that provide the basis for reclassification. *Applied and Environmental Microbiology* 84 (17): e00993-18]

Split of the genus *Lactobacillus*

An evaluation of the taxonomy of *Lactobacillaceae* and *Leuconostocaceae* on the basis of whole genome sequences, core genome phylogeny, pairwise average amino acid identity, clade-specific signature genes, physiological criteria and the ecology of the organisms resulted in a reclassification of the genus *Lactobacillus* into 25 genera (the emended genus *Lactobacillus* included).

Included are organisms that have been referred to as the *Lactobacillus delbrueckii* group, *Paralactobacillus* together with 23 novel genera: *Holzapfelia, Amylolactobacillus, Bombilactobacillus, Companilactobacillus, Lapidilactobacillus, Agrilactobacillus, Schleiferilactobacillus, Loigolactobacilus, Lacticaseibacillus, Latilactobacillus, Dellaglioa, Liquorilactobacillus, Ligilactobacillus, Lactiplantibacillus, Furfurilactobacillus, Paucilactobacillus, Limosilactobacillus, Fructilactobacillus, Acetilactobacillus, Apilactobacillus, Levilactobacillus, Secundilactobacillus* and *Lentilactobacillus.*

The description of the family *Lactobacillaceae* includes all genera that previously have been included in families *Lactobacillaceae* and *Leuconostocaceae.*

The term lactobacilli still designate all organisms that have been classified as *Lactobacillaceae* until 2020.

[Zheng *et al.* (2020). A taxonomic note on the genus *Lactobacillus*: Description of 23 novel genera, emended description of the genus *Lactobacillus* Beijerinck 1901, and union of *Lactobacillaceae* and *Leuconostocaceae*. *Int. J. Syst. Evol. Microbiol.* DOI 10.1099/ijsem.0.004107]

Lactobacillus spp. of probiotic relevance

Lactobacillus delbrueckii

L. delbrueckii is the type species of the genus *Lactobacillus*. Functionally, it is classified as an obligatory homofermentative lactobacilli. Five different subspecies of *L. delbrueckii* have been described:

(*i*) *L. delbrueckii* subsp. *delbrueckii*, which is the type species of the genus *Lactobacillus*,

(*ii*) *L. delbrueckii* subsp. *bulgaricus*, which is included in the traditional starter culture of yoghurt,

(*iii*) *L. delbrueckii* subsp. *indicus*,

(*iv*) *L. delbrueckii* subsp. *lactis*, and

(*v*) *L. delbrueckii* subsp. *sunkii*.

Lactobacillus acidophilus

L. acidophilus has been added to lactic acid fermented milk since the 1930ties. Functionally, *L. acidophilus* is classified as an obligatory homofermentative lactobacilli. *L. acidophilus* is closely related to *Lactobacillus crispatus*.

L. acidophilus grow poorly in milk, and need additional growth factors to grow well, for example, yeast extract.

L. acidophilus is not normally found in spontaneously fermented foods, but is frequently found in the digestive tract of human and animals.

L. acidophilus is a well-known producer of hydrogen peroxide (H_2O_2), and usually lack the enzyme catalase, or other means to convert hydrogen peroxide to water. Due to this inability, hydrogen peroxide is accumulated in the bacterial surroundings. This could be beneficial in some situation, for example by fighting off *E. coli* on the vaginal mucosa, but it can also be a negative factor in critically ill patients at risk for an over-reacting inflammatory response, for example if *L. acidophilus* is given in high doses as probiotics to that type of patients.

Lactobacillus crispatus

L. crispatus is included in the same phylogenetical lactobacilli-group as *Lactobacillu delbrueckii*, and is closely related to *Lactobacillus acidophilus*.

Functionally, *L. cispatus* is classified as an obligatory homofermentative lactobacilli.

L. crispatus over-produce hydrogen peroxide in the presence of oxygen.

L. crispatus is frequently isolated from human vagina, and frequently found in the digestive tract of breast fed infants.

L. crispatus is not usually associated with fermented food.

Lactobacillus gasseri

L. gasseri is included in the same phylogenetical lactobacilli-group as *Lactobacillus delbrueckii. L. gasseri* is closely related to *Lactobacillus johnsonii*.

Functionally, *L. gasseri* is classified as an obligatory homofermentative lactobacilli.

L. gasseri is frequently isolated from human vagina, but is normally not associated with fermented food.

Lactobacillus johnsonii

L. johnsonii is included in the same phylogenetical lactobacilli-group as *Lactobacillus delbrueckii*, and closely related to *Lactobacillus gasseri*.

Functionally, *L. johnsonii* is classified as an obligatory homofermentative lactobacilli.

L. johnsonii produces hydrogen peroxide when exposed to oxygen.

L. johnsonii is often a dominating part of the resident microbiota in vagina, but can also be found in the digestive tract of humans and animals.

L. johnsonii is not usually found in spontaneously fermented foods.

The strain *Lactobacillus johnsonii* La1 (= NCC533; Nestlé) has been used in the probiotic product LC1 (Nestlé). This strain was originally identified as *Lactobacillus acidophilus.*

Ligilactobacillus salivarius

Ligilactobacillus salivarius, former "*Lactobacillus salivarius*" is traditionally regarded as obligatory homofermentative. However, the species contains both homofermentative and facultatively heterofermentative strains. Most strains produce L-lactic acid from available hexoses by homofermentation.

Heterofermentative end-products (lactic acid, acetic acid and ethanol) are produced from hexoses by some strains can also ferment ribose.

L. salivarius is frequently isolated from the digestive tract of mammalians.

A well characterized strain with probiotic impact is *L. salivarius* UCC118 (UCC = University College Cork, Ireland). This particular strain is facultatively heterofermentative.

Lacticaseibacillus casei

Lacticaseibacillus casei, former "*Lactobacillus casei*" (casei means "of cheese") was described and given its name 1916 by the Danish bacteriologist, Orla-Jensen.

Unfortunately, the original type strain was lost and a new type strain (a so called neotype strain) was selected by Hansen and Lessel 1971. However, the neotype strain of *L. casei* had relatively low similarity to most *L. casei*-like isolates from, for example, cheese. Because of this, a new species named "*Lactobacillus paracasei*" was launched, with a description that was based on atypical *L. casei* strains isolated from food.

Some scientist objected to the description of *L. paracasei* and argued for keeping the original classification of *L. casei* (including several subspecies), and they assigned a new neotype strain for *L. casei* that was more similar to environmental isolates than the neotype strain of Hansen and Lessel from 1971.

This proposal was later turned down by the Judicial Commission of the International Committee on Systematics of Bacteria, who decided that the species *L. paracasei* should remain (2008). The consequence of this is that the majority of strains, before labelled *L. casei*, today should be named *L. paracasei*.

True *L. casei* strains seem to be relatively rare in the environment, at least in food, and they are also rare in the digestive tract.

L. casei is functionally classified as facultatively heterofermentative.

L. casei is phylogenetically related to both *Lacticaseibacillus paracasei* and *Lacticaseibacillus rhamnosus.*

Compared with many other lactobacilli species, the *Lacticaseibacillus* spp. have a relatively high proteolytic activity.

True *L. casei*-strains can harbour the gene for catalase production which suggest that *L. casei* can thrive reasonably well in aerobic environments. However, the ability of a lactobacilli taxum to produce catalase is odd, because the lack of catalase has traditionally been a key-feature for the identification of lactobacilli.

The enzyme catalase catalyses the break-down of hydrogen peroxide into oxygen and water.

Lacticaseibacillus paracasei

Lacticaseibacillus paracasei, former "*Lactobacillus paracasei*" grow spontaneously, and to high concentrations in lactic acid fermented dairy products.

L. paracasei will mostly constitute the dominating microbiota in cheese, especially in semi hard cheese made from pasteurised milk where it will reach 10^6 – 10^8 CFU per g cheese. *L. paracasei* are also frequently found in the human digestive tract.

L. paracasei is functionally classified as facultatively heterofermentative, and belongs phylogenetically to the lactobacilli that before 2020 was called the *L. casei*-group.

L. paracasei differs from many other lactobacilli in the ability to (*i*) out compete other organisms in ripening cheese during the ripening period; (*ii*) relative high resistance to heat, and (*iii*) comparably high proteolytic activity.

The probiotic strain "*L. casei* Shirota" is included in the flavoured, sour-milk product Yakult. Yakult is marketed over the whole world by the Japanese company with the same name. Strain Shirota was isolated 1930 by the Japanese scientist Minoru Shirota.

The strain "*L. casei* Shirota" is systematically designated *Lacticaseibacillus paracasei.*

The probiotic strain "*L. casei* defencis" (= DN-114001) that is included in the yoghurt-like product Actimel (Danone) should be designated *Lacticaseibacillus paracasei.*

Other examples on *L. paracasei* strains used commercially as probiotics are *Lacticaseibacillus paracasei* F19 (Arla), and *Lacticaseibacillus paracasei* 8700:2 (= DSM 13434; Probi AB).

Lacticaseibacillus rhamnosus

Lacticaseibacillus rhamnosus former "*Lactobacillus rhamnosus*" was originally classified and named *L. casei* subsp. *rhamnosus* (1968), but has later been reclassified and named *L. rhamnosus* (1989).

L. rhamnosus is facultatively hetero-fermentative.

L. rhamnosus grow spontaneously and rapidly during lactic acid fermentation of milk, and are frequently found in the human digestive tract; especially in breast fed babies where it can reach substantial numbers.

Some strains have now and then been accused for causing infections, but the evidence for this are weak.

L. rhamnosus has been used as probiotics since the 1970ies. Especially one strain, *Lacticaseibacillus rhamnosus* GG (Valio) has been used a lot as probiotics, all over the world.

GG stands for Gorbach and Goldin which were the two scientists that originally isolated and patented the strain. In the beginning, this strain was wrongly identified as *Lactobacillus acidophilus.*

L. rhamnosus GG differ from the majorities of *L. rhamnosus* strains isolated from milk by lacking the ability to ferment lactose.

Lactiplantibacillus plantarum

Lactiplantibacillus plantarum former *Lactobacillus plantarum* (plantarum means "of plants") is phylogenetically connected to *Lactiplantibacillus pentosus* and *Lactiplantibacillus paraplantarum.*

L. plantarum is functionally classified as facultatively heterofermentative.

L. plantarum occur frequently and spontaneously in high numbers in most lactic acid fermented foods. This is especially true when the food is based on plant material, as for example, in brined olives, capers (caper berries), sauerkraut, salted gherkins, kimchi and sour-dough. *L. plantarum* are also frequently found in grape juice and wine.

L. plantarum is frequently isolated from the mucosa of the human digestive tract, from the mouth to the rectum.

L. plantarum have a high tolerance to low pH. The fact that *L. plantarum* frequently predominate in spontaneously, lactic acid fermented food where the pH usually is below 4.0, and also survive the passage through the acid conditions of the human stomach, points to its high resistance to acid conditions.

L. plantarum can ferment many different carbohydrates and has a relatively large genome, indicating that *L. plantarum* has abilities to adapt to many different environments, under different circumstances.

L. plantarum has a growth requirement for manganese and can accumulate high intercellular levels of manganese. Manganese provides a defence for *L. plantarum* against oxygen toxicity by the reduction of oxygen radicals to hydrogen peroxide. The hydrogen peroxide (H_2O_2) can then be converted to oxygen (O_2) and water by manganese co-factored pseudocatalase.

L. plantarum is able to brake-up tannins, and can metabolise flavonoids and phenolic acids.

A well-known probiotic strain, both scientifically and commercially is *Lactiplantibacillus plantarum* 299v (=DSM 9843; Probi AB); a strain originally isolated from human intestinal mucosa.

Other scientifically well characterized probiotic strains are *L. plantarum* WCFS1 (= NCIMB 8826) and *L. plantarum* HEAL9 (=39D; =DSM 15312).

Lactiplantibacillus pentosus

Lactiplantibacillus pentosus, former "*Lactobacillus pentosus*" is closely related to *Lactiplantibacillus plantarum.*

L. pentosus is facultively heterofermentative, and often isolated from the same type of fermented foods as *L. plantarum.*

L. pentosus is able to brake-up tannins, and can metabolise flavonoids and phenolic acids.

Lactiplantibacillus paraplantarum

Lactiplantibacillus paraplantarum, former "*Lactobacillus paraplantarum*" is closely related to *Lactiplantibacillus plantarum*.

L. paraplantarum is facultively heterofermentative, and is often isolated from the same type of fermented foods as *L. plantarum*.

L. paraplantarum is able to brake-up tannins, and can metabolise flavonoids and phenolic acids.

Limosilactobacillus reuteri

Limosilactobacillus reuteri, former "*Lactobacillus reuteri*" was first described 1980, and belongs phylogenetically to a lactobacilli domain that has been called the *L. reuteri*-group; a group that also includes, for example *Limosilactobacillus vaginalis*, *Limosilactobacillus mucosae* and *Limosilactobacillus fermentum*.

L. reuteri are obligatory heterofermentative lactobacilli.

Many *L. reuteri* strains produce the anti-microbial substance, reuterin (b-hydroxy-propionaldehyde) which is an inter-mediate in the glycerol dissimilation.

[Axelsson *et al.* (1989). Production of a broad spectrum antimicrobial substance by *Lactobacillus reuteri*. *Microbial Ecology in Health and Disease*. 2:131–136]

L. reuteri does not normally grow spontaneously in lactic acid fermented foods, and has relatively weak storage stability, compared with *L. plantarum*. The weaker storage stability is presumably due to lower oxygen tolerance.

L. reuteri is frequently found in the digestive tract of humans and animals.

Different strains of *L. reuteri* are marketed as probiotics, primarily by the company BioGaia AB (Sweden). The most prominent strain is *Limosilactobacillus reuteri* SD2112 (=ATCC 55730; = DSM 17938).

Examples of other strains are *L. reuteri* ATCC PTA 6475 and *L. reuteri* ATCC PTA 5289.

Limosilactobacillus fermentum

Limosilactobacillus fermentum, former "*Lactobacillus fermentum*" was first described by Beijerink 1901. Phylogenetically, *L. fermentum* belongs to the *L. reuteri*-group in the family *Lactobacillaceae*, but is within the *L. reuteri* group only distantly related to *Limosilactobacillus reuteri.*

L. fermentum is an obligatory heterofermentative lactobacilli.

L. fermentum is frequently found in high numbers in spontaneously fermented foods, and in the digestive tract of humans and animals.

Examples of commercialized, strains of *L. fermentum* with probiotic claims are *Limosilactobacillus fermentum* PCC (Probiomics Ltd, Australia), and *Limosilactobacillus fermentum* ME-3 (University of Turku, Estonia).

The genus Bifidobacterium

The genus *Bifidobacterium* belongs to the family *Bifidobacteriaceae*, which is included in the order *Bifidobacteriales* that in turn is included in the class Actinobacteria (CN: *Actinomycetes*), under the phylum Actinobacteria (CN: *Actinomycetota*). The genus *Bifidobacterium* contains more than 100 described species.

Bifidobacterium are very different from *Lactobacillaceae* as is seen by the fact that *Bifidobacterium* and *Lactobacillaceae* belong to different phyla. In contrast to *Lactobacillaceae, Bifidobacterium* can grow without vitamins and growth-factors and use inorganic nitrogen (ammonia) as the sole nitrogen source.

Bifidobacterium-cells appear as irregular ("bifid") V or Y-shaped rods with rudimentary branching. They are gram-positive and can ferment sugars to lactic acid and acetic acid following a pathway that is unique for *Bifidobacterium.*

Bifidobacteria are found in the human digestive tract, and are sometimes present in really high numbers in the human colon, and this particularly in breast fed babies. In some individuals bifidobacteria dominate the vaginal microbiota.

Spontaneously fermented foods do not seem to be an optimal environment for bifidobacteria, and normally bifidobacteria are not found in high numbers in fermented foods, at least not the *Bifidobacterium* spp. typically found in the digestive tract.

Bifidobacteria have an optimum pH for growth at 6.5 – 7.0, and don't usually grow below pH 4.5.

Bifidobacteria assimilate inorganic nitrogen (ammonium) for growth and biosynthesis, and are normally sensitive to oxygen.

It has been suggested that bifidobacteria can play a role in the amino acid metabolism of the digestive tract, as bifidobacteria can overproduce different amino acids; an ability that differs between different species, but also between different strains of the same species.

Bifidobacteria are mostly difficult to keep alive in commercial products, and the viable count is sometimes low in products claiming a content of live bifidobacteria.

Anaerobic bacteria as bifidobacteria lack enzymes as superoxide dismutase, catalase and peroxidase, and are also, in most cases unable to perform aerobic metabolism.

Bifidobacteria of probiotic relevance

Bifidobacterium bifidum

Bifidobacterium bifidum is the type species of the genus *Bifidobacterium. B. bifidum* is included in some probiotic products, especially in mixtures containing multiple probiotic strains of different species and genera.

One example of a commercial strain is *B. bifidum* MF 20/5 (Merck).

Bifidobacterium longum

Bifidobacterium longum is perhaps the most frequently occurring *Bifidobacterium* sp. in the human digestive tract, and in vagina.

Specific strains of *B. longum* is used as probiotics.

B. longum was 2008 divided into three subspecies: *Bifidobacterium longum* subsp. *longum*, *Bifidobacterium longum* subsp. *infantis* and *Bifidobacterium longum* subsp. *suis*, and it is *B. longum* subsp. *infantis* that are most prevalent in connection to the human digestive tract, and then especially in that of breast-fed infants.

In commercial contexts and in scientific literature published before 2008, *B. longum* subsp. *infantis* is designated as *Bifidobacterium infantis*.

One examples of a commercial strain is *B. infantis* 35624 (= Bifantis; Procter & Gamble; USA).

Bifidobacterium breve

Bifidobacterium breve can be found in high numbers in faeces of breastfed infants. *B. breve* has also been found, in the intestines of poultry.

B. breve has in clinical trial been given as single-strain probiotics, but more often it has been included as one component in mixtures containing multiple probiotic strains of different species and genera; and sometimes also together with dietary fibres with claimed prebiotic effects.

One example of a commercial strain is *Bifidobacterium breve* Yakult (= BBG-01).

Bifidobacterium animalis

Bifidobacterium animalis has primarily been isolated from faeces of different animals, for example rats and poultry, and from sewage.

B. animalis is divided into the subspecies *B. animalis* subsp. *animails* and *B. animalis* subsp. *lactis*.

In commercial contexts *B. animalis* subsp. *lactis* is mostly called by the older designation "*Bifidobacterium lactis*".

B. animalis seem to be a much hardier species than many other *Bifidobacterium* spp.; *B. animalis* are more resistant to exposure of oxygen.

One commercial strain that has been in use as probiotics for long time is *Bifidobacterium animalis* subsp. *lactis* BB12.

Other taxa with probiotic potential

Faecalibacterium

The genus *Faecalibacterium* with the type species, *Faecalibacterium prausnitzii*, is a gram-positive, rod-shaped, non-spore forming, non-motile anaerobic bacterium.

It is often abundant in the human digestive tract.

F. prausnitzii efficiently produces butyrate and other short-chain fatty acids through the fermentation of dietary fibres.

F. prausnitzii lack known pathogenic potential.

Faecalibacterium prausnitzii belongs to the phylum Firmicutes (CN: *Bacillota*), the class *Clostridia*, the order *Eubacteriales*, and the family *Oscillospiraceae*.

Clostridium butyricum

The species *Clostridium butyricum* belongs to the phylum Firmicutes (CN: *Bacillota*), the class *Clostridia*, the order *Eubacteriales*, and the family *Clostridiaceae*.

C. butyricum is an anaerobic, endospore-forming, gram-positive bacterium.

C. butyricum is, as the name implies an efficient producer of butyric acid.

C. butyricum is uncommonly reported to be involved in clinical situations.

C. butyricum is used as probiotics in Japan, Korea and China.

Bacillus coagulans

The species *Bacillus coagulans* belongs to the phylum Firmicutes (CN: *Bacillota*), the class *Bacilli*, the order *Bacillales*, and the family *Bacillaceae*.

B. coagulans is a lactic acid–producing species. It is a gram-positive, catalase-positive, spore-forming, motile, facultative anaerobe rod.

B. coagulans has been added by the EFSA to their Qualified Presumption of Safety list, and has been approved for veterinary purposes as GRAS by the U.S. Food and Drug Administration's Center for Veterinary Medicine, as well as by the European Union, and is listed by AAFCO for use as a direct-fed microbial in livestock production.

Bacillus subtilis

The species *Bacillus subtilis* belongs to the phylum Firmicutes (CN: *Bacillota*), the class *Bacilli*, the order *Bacillales*, and the family *Bacillaceae*.

B. subtilis is gram-positive, rod-shaped and catalase-positive. It can be regarded as a facultative anaerobe, and it is motile and has flagella.

B. subtilis forms endospores, of extreme resistance to dry heat.

B. subtilis is commonly found in soil, and can form biofilms on plant roots, and occurs on plant surfaces.

B. subtilis can be found in the human digestive tract, and also in the gut microbiota of honey bees.

B. subtilis is not connected to any human diseases, and is not regarded to be infectious.

B. subtilis is spontaneous fermenting, for example, soy beans in sveral traditional foods in East Asia and India. It is used as starter culture in the Japanese fermented food Natto, and claimed to have probiotic abilities.

B. subtilis MB40 (=ATCC PTA-122264; Shakopee, MN, USA), is a commercial strain claimed to have probiotic effects.

Lactic acid bacteria

Which taxa are lactic acid bacteria?

Organisms performing the conversion of carbohydrates to carboxylic acids in lactic acid fermented foods are by tradition called Lactic Acid Bacteria (LAB). The term was used early by food microbiologists, and in 1919, the Danish bacteriologist Orla-Jensen tried to define key features for LAB. He said *"the true lactic acid bacteria form a large natural group of non-motile, non-spore-formers, Gram-positive cocci and rods that at fermentation of sugar mainly produce lactic acid"*.

Followers have added new key features to the list and these criteria are frequently cited in older textbooks. Thus, a typical list of determinative criteria for the description of LAB are that they should be:
(*i*) gram-positive,
(*ii*) able to ferment sugars to mainly lactic acid,
(*iii*) catalase negative,
(*iv*) non-motile,
(*v*) unable to form endospores (endospore = a spore within the cell),
(*vi*) able to grow at low pH (<5.5), and
(*vii*) non-pathogenic.

It must be stressed that these key-criteria are of poor taxonomic relevance as LAB isn't a "natural group" as Orla Jensen assumed, that is, LAB do not form a taxonomic unit based on phylogeny. Therefore, it is to be preferred to refer to the actual genera and/or species in question when dealing with LAB.

Unfortunately, based on old fashioned and artificial taxonomy, different systematically defined taxa that lacking the functionality defining LAB, have been included into LAB.

It is important to remember that LAB is not a systematically defined group based on evolutionary relationships. It is only a term, historically used by food microbiologist when they referred to bacteria responsible for the fermentation of lactic acid fermented foods. Thus, LAB is a functional group and applies to bacteria that are harmless to both food quality and human health, and that occurs spontaneously, and in high numbers in traditional lactic acid fermented foods.

From the systematic point of view, LAB means a relatively wide variety of bacterial taxa. Which genera and species that should be included into the LAB-concept depend much on the variety of foods that are taken into account, and how strictly the quality requirements have been set for the included product. For example, the higher the eating quality is for a lactic acid fermented food product, the fewer types of different types of bacteria are generally involved in the final fermentation.

In a product of poorer quality, all types of more or less unwanted organisms can be found in fairly high numbers in the final product, but those organisms should not be regarded as LAB just because they happen to contaminate a lactic acid fermented product.

Lactic acid bacteria (LAB) =
bacteria that spontaneously dominate the microbiota in traditional, lactic acid fermented foods

The only absolute condition for the organisms involved in lactic acid fermentation of food must be that it produces lactic acid, and that it is harmless to consume in high numbers, and this also for consumers with underlying sicknesses that weaken their immunological defence.

Bacterial taxa frequently occurring in high numbers in traditional, and spontaneously fermented lactic acid fermented foods, are the genera of the family *Lactobacillaceae*, species of the genus *Lactococcus*, and the species *Streptococcus thermophilus* (and closely related species to *S. thermophilus*).

The members of the family *Lactobacillaceae* might all be included in the trivial, often used expression, **lactobacilli**.

It should be held in mind that *Lactococcus* and *Streptococcus thermophilus* from a phylogenetic point of view, not have much in common with *Lactobacillaceae*, other than being members of the same general branch of evolution, that is, the phylum Firmicutes (CN: *Bacillota*).

Bifidobacterium, *Enterococcus* and *Carnobacterium* are examples of genera now and then wrongly referred to as LAB. *Bifidobacterium* is a health promoting genus and are used as probiotics, but is usually not spontaneously dominating in traditional lactic acid fermented foods.

In contrast, *Enterococcus* and *Carnobacterium* are dubious from the health perspective and should be avoided in all sorts of food.

Lactococcus

Species of the genus *Lactococcus* ferment carbohydrates to lactic acid. But in contrast to lactobacilli, *Lactococcus* possesses superoxide-dismutase and menaquinone, that is, systems for efficient utilization of oxygen in the metabolism. *Lactococcus* and particularly the species ***Lactococcus lactis*** is traditionally used as starter culture in the Swedish sour milk product ”filmjölk”, and in the production of most types of European cheeses.

Lactococcus spp. are also used as probiotics, especially in connection with fermented milk-products.

Streptococcus

The genus *Streptococcus* includes a multitude of species, and several species of *Streptococcus* are regarded as pathogenic. The worst pathogen is *Streptococcus pyogenes* which is a "Group A streptococci" (GAS). *S. pyogenes* is a notorious pathogen and cause septic throat, scarlet fever and other pyogenic and septicemic infections.

The infective dose is low, and the incubation period is 1–3 days. Symptoms are sore throat, pain on swallowing, tonsillitis, high fever, headache, nausea, vomiting, malaise and rhinorrhea.

In contrast, many other species of *Streptococcus* are part of the normal microbiota on to the mucous membranes in the mouth and in the digestive tract of both humans and animals.

Streptococcus thermophilus and those *Streptococcus* spp. closely related to *S. thermophilus* are non-pathogenic and are regarded as LAB. *S. thermophilus* ferment carbohydrates to lactic acid, and is traditionally used as starter culture for yoghurt. *S. thermophilus* grow rapidly in milk at temperatures around 45°C.

In nature, *S. thermophilus* is mostly associated to mucous membranes of animals. The high growth temperature of *S. thermophilus* could give a hint that they are more connected to birds than to humans, as birds generally has a relatively high body temperature (around 40°C).

S. thermophilus form together with some other *Streptococcus* spp. a phylogenetic branch of their own (a taxonomic unit) that only distantly is related to other *Streptococcus* spp.

Leuconostoc

The genus *Leuconostoc*, is a member of the family *Lactobacillaceae*, and includes many different species.

Typically, cells have a coccoid morphology, even if they occasionally can be more rod-like.

Leuconostoc ferment carbohydrates to lactic acid, and always with a heterofermentative product pattern, that is, they ferment glucose to lactic acid, ethanol, carbon dioxide, and depending on circumstances, to acetic acid.

Leuconostoc is spontaneously occurring in high numbers in lactic acid fermented milk, and is often dominating the initial fermentation of plant material.

In nature, *Leuconostoc* seems to be associated to insects and snails.

The species *Leuconostoc mesenteroides* has been isolated from human intestinal mucosa.

Weissella

Weissella is a genus in the family *Lactobacillaceae*. Cells of the *Weissella* appear in the microscope as both rods and cocci. *Weissella* ferment carbohydrates to lactic acid, and always with a heterofermentative product pattern.

Weissella is spontaneously occurring in high numbers in lactic acid fermented foods, and can also be found in the human digestive tract.

Oenococcus

Oenococcus is a genus in the family *Lactobacillaceae*, and includes just a few described species. *Oenococcus* appear in the microscope as coccoid or ellipsoidal cells.

Oenococcus ferment carbohydrates to lactic acid, and always with a heterofermentative product pattern.

Oenococcus is spontaneously occurring in high numbers in wine and beer.

Oenococcus can grow fairly well at low pH and in the presence of 10% ethanol.

Pediococcus

Pediococcus is a genus in the family *Lactobacillaceae*, and includes several species. Phylogentically, *Pediococcus* are relatively closely related to *Lactiplantibacillus*.

Cells of *Pediococcus* appear under the microscope as cocci in pair, or in tetrahedrons.

Pediococcus ferment carbohydrates to lactic acid, but can be regarded as facultatively heterofermentative lactobacilli.

Pediococcus is commercially used as starter culture in, for example, smoked, fermented sausages.

Pediococcus is often isolated from fermented vegetables and cereals, and from beer.

Enterococcus

The genus *Enterococcus* includes a lot of species. *Enterococcus* are gram-positive cocci, and are surprisingly resistant to environmental factors as drying and heat.

Enterococci is often detectable in faeces, and can occur spontaneously in different kinds of foods, for example, in cheese, and now and then non-pathogenic strains of *Enterococcus* are even included on purpose in sour-milk products and cheese with claims of being probiotic or beneficial for the taste.

Beware, enterococci are frequently causing urinary tract infections, and enterococci is frequently causing sepsis in hospitalised patients.

Enterococci are easily acquiring antibiotic resistance.

Even milder forms of food poisoning may occur if enterococci are allowed to multitude in high numbers in food. Under these circumstances some enterococci can produce enterotoxins.

Carnobacterium

The genus *Carnobacterium* was formed 1987 after the discovery that some *Lactobacillus* species (former *"Lactobacillus carnis"*; *"Lactobacillus piscicola"* and *"Lactobacillus divergens"*) didn't belong to the genus *Lactobacillus*, or the family *Lactobacillaceae.*

Carnobacterium-cells are rod formed, and ferments carbohydrates to lactic acid, but are unable to grow at pH 5.5 or lower.

Carnobacterium is frequently found in high numbers on meat, or meat products, stored vacuum-packed or packed in modified atmosphere.

Beware, *Carnobacterium* and *Enterococcus* are phylogenetically related.

Carnobacterium can produce biogenic amines.

Pathogens

Enterobacteriaceae

The family *Enterobacteriaceae*, in the order *Enterobacterales*, the class *Gammaproteobacteria* and the phylum Proteobacteria (CN: *Pseudomonadota*), is made up of around 30 genera and 80 species, many of which are potent pathogens including *Escherichia coli*, *Klebsiella pneumoniae*, and *Salmonella enterica*.

Most of the species/genera within the *Enterobacteriaceae* can be assigned to 6 different subfamilies (clades) which are designated as the "*Escherichia* clade", "*Klebsiella* clade", "*Enterobacter* clade", "*Kosakonia* clade", "*Cronobacter* clade" and "*Cedecea* clade".

[Alnajar and Gupta (2017). Phylogenomics and comparative genomic studies delineate six main clades within the family *Enterobacteriaceae* and support the reclassification of several polyphyletic members of the family. *Infection, Genetics and Evolution* 54 (2017) 108–127]

Species of *Enterobacteriaceae* occupy a diverse set of niches in nature, and are associated with a range of diseases, epidemics and infections, involving humans, animals and plants.

All *Enterobacteriaceae* taxa (as all Proteobacteria; CN: *Pseudomonadota*) have a gram-negative cell-wall with lipopolysaccharides (LPS) in the outer cell-membranes.

The genus *Escherichia*, and especially the species *Escherichia coli* are frequently found, and sometimes in abundance in the digestive tract of humans and animals. LPS from *E. coli* have a strong inflammatory-inducing effect on the human immune system.

The opportunistic pathogenic genera *Enterobacter*, *Serratia*, and *Klebsiella* are also often isolated from the human digestive tract.

Clostridium difficile

Clostridium difficile (CN: *Clostridioides difficile*) is an anaerobic rod-shaped bacterium in the Firmicutes (CN: *Bacillota*) phylum, and is the major cause of antibiotic associated diarrhoea.

Due to the fact that the taxonomy of bacteria is a scientific discipline, the division of bacterial taxa are changing with new scientific achievements, and the systematic naming must be altered accordingly. The anaerobic genus *Clostridium* has up to recently included a vast number of different species, but has now been restricted to the type species *Clostridium butyricum* and related species.

[Lawson and Rainey (2016). Proposal to restrict the genus *Clostridium* Prazmowski to *Clostridium butyricum* and related species. *International Journal of Systematic and Evolutionary Microbiology* 66:1009–1016; DOI 10.1099/ijsem.0.000824]

Hence, *C. difficile* is not related to *Clostridium butyricum*, and *C. difficile* is not any longer included in the genus *Clostridium*. Accordingly, *Clostridium difficile* has been reclassified and got a new genus-name, that is, *Clostridioides difficile* and belongs to the family *Peptostreptococcaceae*.

Growth

Requirements

Requirements for growth and multiplication of microorganisms are:

(*i*) the presence of viable microorganisms;
(*ii*) the presence of an energy source;
(*iii*) the presence of nutrients to provide the essential materials from which the microbial cells can be synthesized;
(*iv*) the absence of growth inhibitors;
(*v*) a suitable physical and chemical environment.

Microorganisms can be **heterotrophic**, that is, they obtain their cell carbon from organic compounds, or they can be **autotrophic**, that is, they obtain their cell carbon from carbon dioxide, for example, photosynthesizing organisms.

Most of the bacterial taxa associated to humans (and animals) are heterotrophic, and extract energy from oxidation–reduction reactions, where the electron acceptors in the energy-yielding processes are organic compounds, oxygen, nitrate or sulphate.

Fermentation is an energy-yielding process through oxidation–reduction of organic compounds, taking palace under anaerobic conditions.

Inhibitory factors

Water activity

When a substance is solved in water, the water molecules become oriented in relation to the molecules of the involved substance, which decrease the entropy of the system. And it becomes more difficult for the water molecules to leave the liquid phase and pass over into gas phase, that is, the vapour pressure of water over the liquid becomes lower. The relation between concentration and vapour pressure of an ideal solution is regulated by Raoul's law:

$$p/p_o = n_2/n_1 + n_2$$

where p and p_o are the vapour pressure over the solution and the solvent, respectively. And n_2 and n_1 is the number of molecules of solved substance and the number of molecules of the solvent, respectively. The quota between p and p_o defines the water activity (a_w).

$$a_w = p/p_o$$

The water activity over pure water is 1.0, and the relative humidity in the gas phase (RH in per cent) is equal to a_w x 100.

The lowest a_w–limit of growth for some different bacterial taxa:
0.97 for *Pseudomonas*;
0.95 for *Enterobacteriaceae*;
0.93 for *Lactobacillaceae*;
0.85 for *Staphylococcus*.

For bacteria in general, the minimum water activity for growth in food is mostly 0.93 (with extremes going down to 0.85).

The lowest water activity for growth of yeast is 0.85 (with extremes, down to 0.80), and for moulds 0.75 (with extremes, down to 0.60).

Natrium chloride is a well proved means to decrease the growth and multiplication of microorganism in food. The inhibitory effect is mainly due to the decreased water activity, but added to the decreased water activity is a minor inhibitory effect added caused by the salt itself.

pH

The hydrogen ion concentration (pH) of the environment surrounding microorganisms will affect the nature of the cell surface through the dissociation of acids and bases. This will affect surface properties of the cell. To some extent the cell-membranes will protect the interior of the cell, but nevertheless the cell metabolism is readily influenced by environmental pH.

The lowest pH-limit for growth of most bacteria is pH 3.0; while the corresponding pH-limit for moulds and yeasts is 1.4.

The susceptibility for low pH of some different bacterial taxa decreases in the order:
> *Pseudomonas* (most sensitive to low pH)
> *Enterobacteriaceae*
> *Bacillus*
> *Enterococcus*
> *Leuconostoc*
> *Pediococcus*
> *Lactoplantibacillus* (highest resistance against low pH).

Organic acids possess growth inhibitory effects against microorganisms. Generally, their inhibitory effect is increasing at decreasing pH. When the acid is appearing in un-dissociated form at pH below the pK_a-value (the pH where 50% of the substance is ionized) the acids become more growth-inhibitory. The un-dissociated acid molecule can be up to four times as inhibitory as the dissociated molecule.

Examples of acids allowed by the Swedish authorities in food are L-ascorbic acid, benzoic acid, citric acid, lactic acid, propionic acid, sorbic acid, malic acid and acidic acid.

Nutrients

The nutrients required for growth are:

(*i*) the main elements in biomass, that is, carbon, hydrogen, oxygen and nitrogen;

(*ii*) elements of phosphorous, potassium, sulphur and magnesium;

(*iii*) vitamins and other organic growth factors (for some microorganisms);

(*iv*) trace-elements.

All inorganic and organic forms of nitrogen can act as nitrogen source for microorganisms, but the ability to utilize the different forms vary between differs sorts of microorganisms.

The nitrogen is metabolized to provide the cell with protein, nucleic acids and cell wall polymers.

Amino acids are for many microorganisms needed as a growth factor.

Phosphorous is mostly utilized as inorganic phosphate. Alternatives are organic phosphates, for example glycerol-phosphates and phosphor-lipids.

Phosphate is incorporated into the nucleic acids, phospholipids and cell wall polymers.

The potassium (K) requirement for growth corresponds to a growth yield of roughly 60 g dry biomass per g potassium. Much of the potassium is incorporated in RNA.

Sulphur is commonly provided in its inorganic forms, usually sulphate, or in the form of cysteine or methionine.

One gram of magnesium is sufficient for synthesizing 300 to 900 g dry biomass, and is inversely proportional to the amount of RNA in the biomass.

A rough estimation of the quantitative requirements for the more important trace elements are, in gram element per 100 g dry biomass, for: Ca, 0.10; Fe, 0.015; Mn, 0.005; Zn, 0.005; Cu, 0.001; Co, 0.001; Mo, 0.001.

Logarithmic cell multiplication

Bacteria multiply by division – one bacterial cell divides and becomes two. The time between two divisions is the **doubling time**. Under good growth conditions the doubling time can be short as 20 minutes.

If it is assumed that the growth conditions for bacteria giving rise a doubling time of 20 min could be kept for 24 hours, then one bacterial cell (weight $5x10^{-13}$ gram) would have multiplied to $4.7x10^{21}$ bacterial cells, weighing 2 400 000 kg.

The growth dynamics of microorganisms (biomass) are as follows: During an infinitely short time-interval (dt) the expected increase in biomass (dx) is proportional to the amount biomass (x) present and to the time interval:

$$dx = \mu x.dt$$

that is:

$$dx/dt = \mu x$$

The differential coefficient (dx/dt) expresses the growth rate of the population. The parameter μ which represents the rate of growth per unit amount of biomass $(1/x)(dx/dt)$, is called the **specific growth rate** and has the dimension of reciprocal time ($1/t$).

If the specific growth rate (μ) is constant, integration of $dx/dt = \mu x$ gives:

$$\ln x = \ln x_0 + \mu t$$

where x_0 is the biomass when $t = 0$. The plot of $\ln x$ against time will be a straight line where the direction coefficient is μ. If the logarithms are converted to the base of 10 then $\ln x = \ln x_0 + \mu t$ becomes:

$$\log x = \mu t/2.30 + \log x_0$$

By putting $\ln x = \ln x_0 + \mu t$ in the form $\ln (x/x_0) = \mu t$
It follows that:

$$x = x_0\, e^{\mu t}$$

Growth which accords with this law is called constant **exponential**, or **logarithmic growth** (Figure 16). The basic measure of the growth rate is the specific growth rate.

$$\mu = 2.30\ (\log x - \log x_0)/t$$

The relation between specific growth rate and the **doubling time** (t_d) of the biomass is obtained from
$\ln (x/x_0) = \mu t$ by putting $x = 2x_0$ and $t = t_d$:

$$t_d = \ln 2/\mu = 0.693/\mu$$

In populations of macro-organism the doubling time is called generation time, and it is calculated as the average time between two consecutive generations in the lineages of a population. The generation time of humans is typically between 22—33 years.

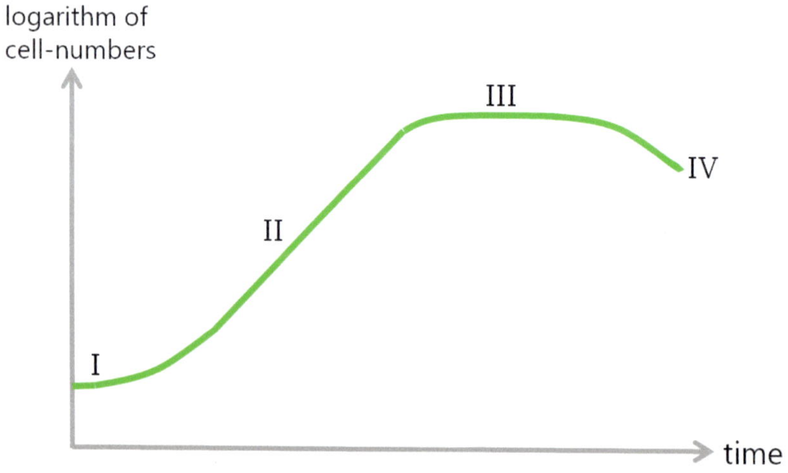

Figure 16. Growth curve for bacteria. The different parts of the curve are called (I) lag phase, (II) logarithmic phase, (III) stationary phase, and (IV) declination phase.

Temperature

The interior of a microbial cell will be the same as the temperature of its surrounding, due to the small size of the microorganism. The tiny size results in a ratio between cell surface and body mass that makes it impossible to uphold an internal cell temperature deviating from the surrounding.

The ambient temperature affects the growth rate. Different microorganisms are differently affected by temperature. Bacteria that have optimal growth at 37°C are called mesophilic, and usually they can multiply relatively well in the temperature interval of 15—40°C.

Bacteria that can multiply at refrigeration temperatures are called psychrotrophic (≤5°C); while bacteria that can't multiply above 20°C and have an optimal growth at 15°C or lower, are called psychrophilic.

Bacteria that have optimal growth at 65–70°C and can't grow at temperatures below 40°C are called thermophilic.

Most bacterial taxa associated to humans and warm-blooded animals are mesophilic.

Bacteria growing on food in the refrigerator are to predominantly psychrotrophes, and can without problem grow and multiply at 0°C, but the growth rate is considerably lower than at, for example, 4°C (Figure 17). The spoilage microbiota on cold stored meat is dominated by species as *Pseudomonas fragi*, and *Pseudomonas lundensis*, wherever in the world the refrigerator is situated.

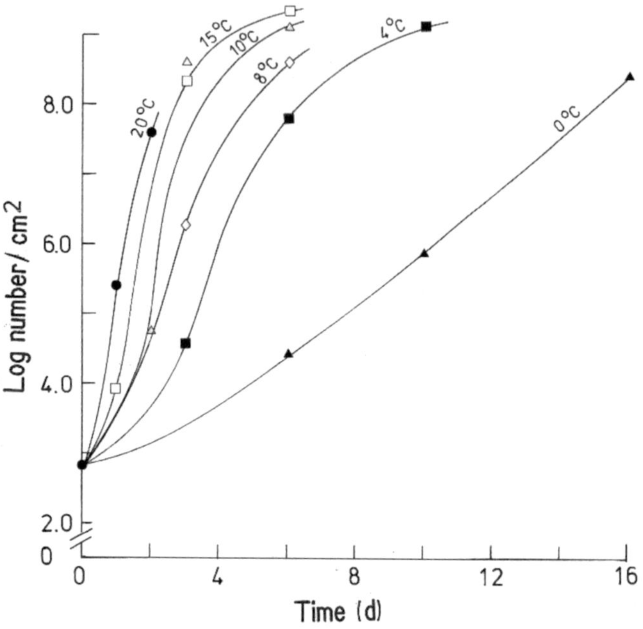

Figure 17. Viable count of spoilage bacteria (mostly *Pseudomonas* spp.) growing on pork loins stored at different temperatures.

The meat starts to give off a foul smell when the number of bacteria reach 10^7 CFU per cm^2 meat surface. As can be seen in Figure 17; the time to reach this level is highly dependent on storage temperature.

Death

In some places, and in some situations live bacteria are highly unwanted, and only stone-dead bacteria can be accepted. The crucial question then become: when are they all dead?

There are several options available to kill bacteria. Bacteria can be killed by heat which probably is the safest and most reliable method, but they can also be efficiently killed by radioactive radiation, for example gamma-radiation, and by chemical oxidation, for example with ethylene-oxide.

If the task is to kill all bacteria in a substrate intended for use as, for example, culturing media for production of a probiotic strain: A question could arise – to what temperature should then the substrate be heated – what temperature can guarantee that absolutely all bacteria are dead?

Put in that way, a true answer is – no single temperature can give theoretical guarantees for absolutely absence of live bacteria. The question must first be specified with the number of bacteria present in the substrate to be heated.

Theoretically, single-cell microorganisms are heat-killed after the same dynamics as a radioactive decay (nuclear disintegration), that is, a first-order process in which the rate is dependent upon the concentration (or amount) of only one component.

In other words, the microorganisms are heat inactivated according to a geometric regression where the same fraction of remaining cells are inactivated in each successive time-interval. The process is described by the equation:

$$\log N_T = -T/D + \log N_0$$

Where N_0 is the starting number of live microorganisms; **D** is the inactivation rate, that is, the time it takes to reduce the population with 90% at a given temperature. And N_T is the number of living microorganisms after the heating time T. It follows from this that if the logarithmic number of surviving microorganisms is plotted against time the curve will be a straight line, the so called, inactivation curve (Figure 18).

This is an idealized situation, in practical inactivation experiments, it is common that the heat inactivation kinetics in different way deviates from the straight line, especially in the start and in the later part of the curve; the line has a tendency to level out in the end.

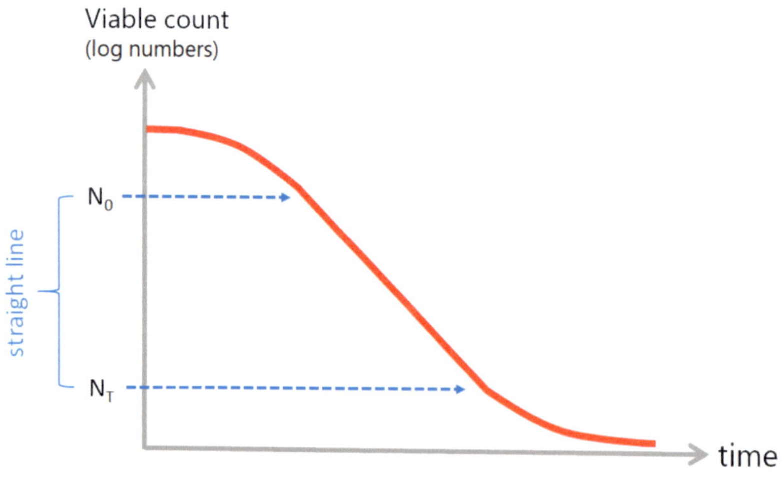

Figure 18. Heat inactivation of microorganisms at constant temperature.

The **D-value** describes the heat-resistance of treated microorganism in the environment where the heat exposure takes place. The D-value can be used for comparison of different type of microorganisms provided they are tested in similar heat environment.

The D-value at 60°C (at 1.0 water activity) for members of the family *Enterobacteriaceae* can be expected to be within the interval 1–10 min, while the corresponding interval for *Lactobacillaceae* is 5–120 min.

The heat resistance of microorganisms is very different if the heating is taken place in moist heat (water activity = 1.0) or dry heat (water activity <1.0). The heat resistance is dramatically increased at lower water activity; and is decreased already at a water activity of 0.98.

To keep the water activity of 1.0 at temperatures above 100°C, requires an increased steam pressure. A commonly used temperature in saturated water steam is 121°C (called autoclavation) which requires a steam overpressure of 1 atm (=101.325 Pa).

Some bacterial taxa in the phylum Firmicutes (CN: *Bacillota*) forms endospores which are dormant, non-metabolizing lifeforms. Endospores are formed within a bacterial cell ("endo" means within), and are extremely resistant to heat and other life-killing agents.

Several fungi produce spores but they are formed outside the metabolizing cells. Fungi-spores are resistant against dehydration and ultraviolet radiation but are not especially heat-resistant.

An example of D-value at 121°C (at saturated water steam) for endospores of *Bacillus stearothermophilus* (CN: *Geobacillus stearothermophilus*) are 5 min, and for *Clostridium sporogenes* 0.5 min.

The heat resistance for both vegetative cells and endospores is dramatically increased in dry heat. Dry heat is, formally a condition where the water activity is below 1.0; often the dramatically change in heat-resistance occurs around the water activity 0.95.

If the D-value for a particular taxum of endospores is 0.5 min at moist heat, the corresponding value at dry heat can be expected to be around 30 min.

Empirically found experiences have shown that the relation between the logarithmic D-values for different temperatures (under equal test-conditions) is a straight line, that is:

$$\log D_T = 1/z\,(T_O - T) + \log D_{T_O}$$

where T_O is the initial temperature, D_{T_O} is specified and T is the new temperature for D_T. The **z-value** represent the increase in temperature needed to change the D-value with a factor of 10. The z-value can be used for forecasting the heat resistance (the D-value) for a certain microorganism (at a specific heating environment) at different temperatures.

One example of how the connection between D-values and the z-values can be used for calculating the theoretical D-value of spores is given below.

Study example: The resistance of endospores

A plot of the thermal resistance of *Bacillus subtilis* var. *niger* spores (log D) against temperature was linear between 37°C and 190°C with a calculated z-value of 23°C. The data indicate that bacterial spores which are treated in dry heat at an environment with a relative humidity near zero are inactivated mainly by a drying process, irrespectively of temperature. By extrapolation of the thermal resistance plot obtained under these conditions, the D-value at 0°C would be about 4 years, that is, it takes 4 years to reduce the spore population with 90%.

If a constant z-value is assumed over the entire temperature range, and extrapolates the inactivation data of the present example, the D-value of *B. subtilis* var. *niger* spores near 0 K would be about 10^{12} years. This suggests that bacterial spores will keep their viability, at least theoretically, for extremely long times in real cold environments, as for example in space.

[Molin (1977). Inactivation of *Bacillus* spores in dry systems at low and high temperatures. *Journal of General Microbiology* 101:227–231]

Evolution

The first scientist to suggest biological evolution was the French zoologist Jean-Baptiste Lamarck. He presented his ideas in "Philosophie zoologique" (1809), but without gaining much hearing for the idea.

Maybe the lack of sympathy was because he failed to give a good explanation for the driving force of evolution. Lamarck proposed that evolution is caused by inherited experience which few believed in at the time, especially as the general idea was that Earth only was 6000 years old (calculated from the Bible), that is, a too tiny time-span to allow an evolutionary development.

All credit for the evolutionary idea went to Charles Darwin that 1859 published the book "On the origin of species by means of natural selection". Geologists had by then proved that the Earth actually was by far older than expected earlier.

Simplified, the theory of Darwin consists of two theses, (*i*) successive (gradual) development and (*ii*) natural selection. The theory of natural selection gives an explanation to the driving force, or the principle of evolution.

The theory of natural selection was later further developed in the so called "Synthetic theory" which postulates that (*i*) individuals in a population have genomic differences, and (*ii*) individuals (genomes) with the most efficient reproduction will come to dominate the population.

Individuals with the most efficient reproduction are not necessarily the best individuals from a human perspective, and the best species is not necessarily the one that momentarily is dominating a specific niche in nature, but more the species that remains over time – the longer the better.

It is important to hold in mind that the theory of evolution is not a "natural law". The theory of evolution cannot be used for broadcasting future evolution. It can only, as best give an explanation to what already has happen. However, the theory of evolution is an established scientific hypothesis with major impact on human thoughts about the biological life.

A key-concept in the theory of evolution is "natural selection". Natural selection was early rephrased by the British liberal philosopher Herbert Spencer (1820 –1903) to "survival of the fittest", when he drew parallels between Darwin's theory of evolution and his own economical ones. This interpretation of natural selection into survival of the fittest has had major impact, not only on the biological thought but also on thoughts of the society. Hence, the driving principle of evolution has often been described in terms of competition and struggle, and even in terms of warfare as battle and fight.

However, is this actually giving a fair description of "natural selection"?

No, said the Russian philosopher Pyotr Alexeyevich Kropotkin (Peter Kropotkin, 1842 –1921), and argues in "Mutual aid: A factor of evolution" that natural selection is more driven by co-operation than by competition.

Let us take a look on reality, that is, the life going on in nature all around us, in this day.

Are different individual life forms (organisms) in a population, different species, or different divisions of life really involved in a competitive battle, on life and death?

Is life a battle where different life forms with all means try to erase each other from existence, in order to take over and dominate, and in this way create natural niches of its own?

Perhaps, it might look that way when a parasitic outburst is killing off a large number of targeted organisms. However, if the parasite in question stays in the environment, the target organism will with time start to tolerate the parasitic organism and the parasite will lose its pathogenic aggressiveness, that is, both parasite and target (host) will change (evolve) into coexistence. The driving force in this "natural selection" is obviously the drift of adaptation.

But what about so called “invading species”, those that outcompete resident species, a phenomenon frequently discussed in media for both plants and animals?

These scenarios can happen when species totally new for a certain environment (biotope) has been trespassing, which can be due to changes in the environmental conditions due to either (*i*) non-biological changes in the environment (for example, changes in temperature or rainfall) or (*ii*) biological changes (introduction of alien organisms to the specific biotope).

Unbalanced biotopes are often a result of human activities, and continuing human activities maintain unbalances, and hence, can give the impression of successful completion of certain types of organisms. However, if the biotope is left to nature, time will equalize the unbalance and old and new species will either adapt or vanish.

Why then say that competition is the key word, and that the most efficient competitor will rule the biotope with time? When reality show something else: A biotope in balance (static environmental conditions) will always aim at as high biological diversity as possible. It is not aiming at singular “survivors”, after a major battle of power for the fittest.

The numbers of examples of mutual adoption and cooperation in biology is as vast as there are organisms: One striking example on individual cooperation is the bee community; furthermore, bees are also in close cooperation with flowering plants.

The individual cells of an animal and a plant are in close cooperation with each other to form the individual creature, or plant.

Mitochondria and the animal cell, or chloroplasts and the plant cell, are most probably the result of former bacterial adoption to eukaryotic cells.

The rhizosphere, that is, the area around living plant-roots, is site of high biological activity, and of high microbial diversity.

The lower part of the gastro-intestinal tract of mammalians has an extremely dense population of bacteria, in an order of magnitude seldom found in any other niches in nature. And provided the host is in good health and the digestion is not disturbed, the bacterial diversity will be equally high. Thus, the competition between the different bacterial taxa ought to be equally high, but nevertheless, the diversity is high and driven towards increasing plurality.

If competition was the actual driving force of biology, the diversity ought to approach singularity instead, but it isn't. And it isn't because competition is the wrong concept in nature, the force of nature – the natural selection – is driven by adoption leading to cooperation with the environment, both in respect of the inorganic environment and the organic environment.

Competition is not a winning concept in nature, which is a fact that should be held in mind when discussing the effect of probiotics in the mouth and in the digestive tract. The probiotic organism is not meant to take over and make a diverse intestinal microbiota to a single-organism niche. Probiotics should counteract some unwanted taxa, but open up for more beneficial taxa. Probiotics should increase the diversity in the digestive tract, not decrease diversity.

Made in the USA
Las Vegas, NV
04 November 2026

57436829R00157